T0321592

HYPERBOLICITY AND SENSITIVE CHAOTIC DYNAMICS AT HOMOCLINIC BIFURCATIONS

Already published

1   W.M.L. Holcombe *Algebraic automata theory*
2   K. Petersen *Ergodic theory*
3   P.T. Johnstone *Stone spaces*
4   W.H. Schikhof *Ultrametric calculus*
5   J.-P. Kahane *Some random series of functions, 2nd edition*
6   H. Cohn *Introduction to the construction of class fields*
7   J. Lambek & P.J. Scott *Introduction to higher-order categorical logic*
8   H. Matsumura *Commutative ring theory*
9   C.B. Thomas *Characteristic classes and the cohomology of finite groups*
10 M. Aschbacher *Finite group theory*
11 J.L. Alperin *Local representation theory*
12 P. Koosis *The logarithmic integral I*
13 A. Pietsch *Eigenvalues and s-numbers*
14 S.J. Patterson *An introduction to the theory of the Riemann zeta-function*
15 H.J. Baues *Algebraic homotopy*
16 V.S. Varadarajan *Introduction to harmonic analysis on semisimple Lie groups*
17 W. Dicks & M. Dunwoody *Groups acting on graphs*
18 L.J. Corwin & F.P. Greenleaf *Representations of nilpotent Lie groups and their applications*
19 R. Fritsch & R. Piccinini *Cellular structures in topology*
20 H Klingen *Introductory lectures on Siegel modular forms*
21 P. Koosis *The logarithmic integral II*
22 M.J. Collins *Representations and characters of finite groups*
24 H. Kunita *Stochastic flows and stochastic differential equations*
25 P. Wojtaszczyk *Banach spaces for analysts*
26 J.E. Gilbert & M.A.M. Murray *Clifford algebras and Dirac operators in harmonic analysis*
27 A. Frohlich & M.J. Taylor *Algebraic number theory*
28 K. Goebel & W.A. Kirk *Topics in metric fixed point theory*
29 J.E. Humphreys *Reflection groups and Coxeter groups*
30 D.J. Benson *Representations and cohomology I*
31 D.J. Benson *Representations and cohomology II*
32 C. Allday & V. Puppe *Cohomological methods in transformation groups*
33 C. Soulé et al *Lectures on Arakelov geometry*
34 A. Ambrosetti & G. Prodi *A primer of nonlinear analysis*
35 J. Palis & F. Takens *Hyperbolicity and sensitive chaotic dynamics at homoclinic bifurcations*
37 Y. Meyer *Wavelets and operators*
38 C. Weibel *An introduction to homological algebra*
39 W. Bruns & J. Herzog *Cohen–Macaulay rings*
40 V. Snaith *Explicit Brauer induction*

HYPERBOLICITY AND SENSITIVE CHAOTIC DYNAMICS AT HOMOCLINIC BIFURCATIONS

Fractal Dimensions and Infinitely Many Attractors

Jacob Palis

Professor of Mathematics, IMPA
Rio de Janeiro

Floris Takens

Professor of Mathematics
University of Groningen

CAMBRIDGE
UNIVERSITY PRESS

CAMBRIDGE UNIVERSITY PRESS
Cambridge, New York, Melbourne, Madrid, Cape Town, Singapore, São Paulo

Cambridge University Press
The Edinburgh Building, Cambridge CB2 2RU, UK

Published in the United States of America by Cambridge University Press, New York

www.cambridge.org
Information on this title: www.cambridge.org/9780521390644

© Cambridge University Press 1993

First published 1993
First paperback edition 1995

Appendix 5 is reproduced with permission of the *Annals of Mathematics*

A catalogue record for this publication is available from the British Library

ISBN-13 978-0-521-39064-4 hardback
ISBN-10 0-521-39064-8 hardback

ISBN-13 978-0-521-47572-3 paperback
ISBN-10 0-521-47572-4 paperback

Transferred to digital printing 2005

CONTENTS

Preface   vii

0 – Hyperbolicity, stability and sensitive chaotic
     dynamical systems   1
    §1   Hyperbolicity and stability   1
    §2   Sensitive chaotic dynamics   8

1 – Examples of homoclinic orbits in dynamical systems   11
    §1   Homoclinic orbits in a deformed linear map   12
    §2   The pendulum   12
    §3   The horseshoe   14
    §4   A homoclinic bifurcation   15
    §5   Concluding remarks   15

2 – Dynamical consequences of a transverse homoclinic intersection   18
    §1   Description of the situation – linearizing coordinates
          and a special domain R 18
    §2   The maximal invariant subset of R – topological analysis   22
    §3   The maximal invariant subset of R – hyperbolicity and
          invariant foliations   23
    §4   The maximal invariant subset of R – structure   27
    §5   Conclusions for the dynamics near a transverse homoclinic orbit   30
    §6   Homoclinic points of periodic orbits   31
    §7   Transverse homoclinic intersections in arbitrary dimensions   31
    §8   Historical note   32

3 – Homoclinic tangencies: cascades of bifurcations,
     scaling and quadratic maps   34
    §1   Cascades of homoclinic tangencies   35
    §2   Saddle-node and period doubling bifurcations   37
    §3   Cascades of period doubling bifurcations and sinks   40
    §4   Homoclinic tangencies, scaling and quadratic maps   45

4 – Cantor sets in dynamics and fractal dimensions   53
    §1   Dynamically defined Cantor sets   53
    §2   Numerical invariants of Cantor sets   60
    §3   Local invariants and continuity   82

5 – Homoclinic bifurcations: fractal dimensions
and measure of bifurcation sets   92

§1   Construction of bifurcating families of diffeomorphisms   93
§2   Homoclinic tangencies with bifurcation set of small relative
measure – statement of the results   99
§3   Homoclinic tangencies with bifurcation set of small relative
measure – idea of proof   102
§4   Heteroclinic cycles and further results on measure of
bifurcation sets   107

6 – Infinitely many sinks and homoclinic tangencies   112
§1   Persistent tangencies   113
§2   The tent map and the logistic map   116
§3   Hénon-like diffeomorphisms   119
§4   Separatrices of saddle points for diffeomorphisms near
a homoclinic tangency   125
§5   Proof of the main result   129
§6   Sensitive chaotic orbits near a homoclinic tangency   131

7 – Overview, conjectures and problems – a theory of
homoclinic bifurcations – strange attractors   132
§1   Homoclinic bifurcations and nonhyperbolic dynamics   133
§2   Strange attractors   138
§3   Summary, further results and problems   148

Appendix 1 – Hyperbolicity: stable manifolds and foliations   154

Appendix 2 – Markov partitions   169

Appendix 3 – On the shape of some strange attractors   178

Appendix 4 – Infinitely many sinks in one-parameter families
of diffeomorphisms   180

Appendix 5 – Hyperbolicity and the creation of homoclinic orbits,
reprinted from *Annals of Mathematics* **125** (1987)   185

References   223
Index   233

PREFACE

Homoclinic bifurcations, which form the main topic of this monograph, belong to the area of dynamical systems, the theory which describes mathematical models of time evolution, like differential equations and maps. Homoclinic evolutions, or orbits, are evolutions for which the state has the same limit both in the "infinite future" and in the "infinite past".

Such homoclinic evolutions, and the associated complexity, were discovered by Poincaré and described in his famous essay on the stability of the solar system around 1890. This associated complexity was intimately related with the breakdown of power series methods, which came to many, and in particular to Poincaré, as a surprise.

The investigations were continued by Birkhoff who showed in 1935 that in general there is near a homoclinic orbit an extremely intricate complex of periodic solutions, mostly with a very high period.

The theory up to this point was quite abstract: though the inspiration came from celestial mechanics, it was not proved that in the solar system homoclinic orbits actually can occur. Another development took place which was much more directed to the investigation of specific equations: in order to model vacuum tube radio receivers Van der Pol introduced in 1920 a class of equations, now named after him, describing nonlinear oscillators, with or without forcing. His interest was mainly in the periodic solutions and their dependence on the forcing. In later investigations of this same type of equations, around 1950, Cartwright, Littlewood and Levinson discovered solutions which were much more complicated than any solution of a differential equation known up to that time.

Now we can easily interpret this as complexity caused by (the suspension of) a horseshoe, which in its turn is a consequence of the existence of one (transverse) homoclinic orbit, but that is inverting the history...

In fact, Smale, who originally had focussed his efforts on gradient and gradient-like dynamical systems, realized, when confronted with these complexities, that he should extend the scope of his investigations. Seventy years after Poincaré, Smale was again shocked by the complexity of homoclinic behaviour! By the mid 1960 he had a very simple geometric example (i.e. no formulas but just a picture and a geometric description), which could be completely analysed and which showed all the complexity found before: the horseshoe. This new prototype dynamical model, the horseshoe, together with the investigations in the behaviour of geodesic flows of manifolds with negative curvature (Hadamard, Anosov), grew, due to the efforts of a number of mathematicians, to an extension of the gradient-like theory which we now know as hyperbolic dynamics, and which in particular provides models for very complex (chaotic) dynamic behaviour.

Around 1980 this hyperbolic theory was used by Levi to reanalyse the qualitative behaviour of the solutions of Van der Pol's equation, largely extending the earlier results. He proved that, besides all the complexity we know from the hyperbolic theory, even the new and extreme complexity associated with homoclinic bifurcations, which we shall consider below, actually exists in the solutions of this equation.

Homoclinic bifurcations, or nontransverse homoclinic orbits, become important when going beyond the hyperbolic theory. In the late 1960s, Newhouse combined homoclinic bifurcations with the complexity already available in the hyperbolic theory to obtain dynamical systems far more complicated than the hyperbolic ones. Ultimately this led to his famous result on the coexistence of infinitely many periodic attractors and was also influential on our own work on hyperbolicity or lack of it near homoclinic bifurcations. These developments form the main topic of the present monograph, of which we shall now outline the content.

We start with Chapter 0 that presents general background information about the hyperbolic theory and its relation to (structural) stability of systems, and discuss as well some initial aspects of chaotic dynamics; many results on stable manifolds and foliations are stated, and their proofs sketched, in Appendix 1. The later chapters, except the last one, do not depend on the results described in this chapter and are basically self-contained.

In Chapter 1, we give a number of simple examples of homoclinic orbits and bifurcations. Chapter 2 discusses the horseshoe example and shows how it is related to homoclinic orbits. Then, in Chapter 3, we consider some preliminary and more elementary consequences of the occurrence of a homoclinic bifurcation, especially in terms of cascades of bifurcations which have to accompany them.

In Chapters 4, 5, and 6, we come to our main topic: the investigation of situations where there is an interplay between homoclinic bifurcations and nontrivial basic sets, the sets being the building blocks of hyperbolic systems with complex behaviour. Since such basic sets often have a fractal structure, we start in Chapter 4 with a discussion of Cantor sets and fractal dimensions like Hausdorff dimension. In Chapter 5 the emphasis is on hyperbolicity near a homoclinic bifurcation associated with a basic set of small Hausdorff dimension. Then, in Chapter 6, we discuss types of homoclinic bifurcations which yields, in a persistent way, complexity beyond hyperbolicity. In this chapter we provide a new, and more geometric, proof of Newhouse's result on the coexistence of infinitely many periodic attractors. Finally, in Chapter 7 we present an overview of recent results, including specially Hénon-like and Lorenz-like strange attractors. We also pose new conjectures and problems which may lead to a better understanding of nonhyperbolic dynamics (the "dark realm" of dynamics) and the role of homoclinic bifurcations.

Summarizing, we deal with the following rather striking collection of dynamical phenomena that take place at the unfolding of a homoclinic tangency

- transversal homoclinic orbits, which in turn are always associated to horseshoes (invariant hyperbolic Cantor sets): Chapters 1 and 2,

- cascades of homoclinic tangencies, i.e. sequences in the parameter line whose corresponding diffeomorphisms exhibit a homoclinic tangency: Chapter 3,

and for families of locally dissipative diffeomorphisms,

- cascades of period doubling bifurcations of periodic attractors (sinks): Chapter 3,

- cascades of critical saddle-node cycles: Chapter 7,

- residual subsets of intervals in the parameter line whose corresponding diffeomorphisms exhibit infinitely many coexisting sinks: Chapter 6 and Appendix 4,

- positive Lebesgue measure sets in the parameter line whose corresponding diffeomorphisms exhibit a Hénon-like strange attractor: Chapter 7 and Appendix 3,

- prevalence of hyperbolicity when the fractal (Hausdorff) dimension of the associated basic hyperbolic set is smaller than 1: Chapters 4 and 5 and Appendix 5,

- nonprevalence of hyperbolicity when the above fractal dimension is bigger than 1: Chapter 7.

In our presentation we mainly restrict ourselves to diffeomorphisms in dimension 2 (which is the proper context to investigate classical equations e.g. the forced Van der Pol equation), although extensions to higher dimensions are mentioned; also we concentrate mainly on the general theory as opposed to the analysis of specific equations. Consequently a number of topics like Silnikov's bifurcations and the Melnikov method are not discussed.

We hope that by putting this material together, rearranging it to some extent and pointing to recent and possible future directions, these results and their proofs will become more accessible, and will find their central place in dynamics which we think they merit.

We wish to thank a number of colleagues from several different institutions as well as Ph. D. students from the Instituto de Matematica Pura e Aplicada

(IMPA) who much helped us in writing this book. Among them we mention M. Benedicks, L. Carleson, M. Carvalho, L. Diaz, P. Duarte, R. Mañé, L. Mora, S. Newhouse, M. J. Pacífico, J. Rocha, D. Ruelle, R. Ures, J. C. Yoccoz and most especially M. Viana. Thanks are also due to Luiz Alberto Santos for his fine typing of this text.

CHAPTER 0

HYPERBOLICITY, STABILITY
AND SENSITIVE CHAOTIC
DYNAMICAL SYSTEMS

In this chapter we give background information and references to the literature concerning basic notions in dynamical systems that play an important role in our study of homoclinic bifurcations. Essentially, the chapter consists of a summary of the hyperbolic theory of dynamical systems and comments on sensitive chaotic dynamics. This is intended both as an introduction to the following chapters and to provide a more global context for the results to be discussed later and in much more detail than the ones presented in this Chapter 0.

In the first section we concentrate on hyperbolicity and emphasize its intimate relation with various forms of (structural) stability. In the second section we discuss several aspects of sensitivity ("chaos") and indicate how it occurs in hyperbolic systems.

§1 Hyperbolicity and stability

These two concepts, hyperbolicity and (structural) stability, have played an important role in the development of the theory of dynamical systems in the last decades: the hyperbolic theory was mostly developed in the 1960's, having as a main initial motivation the construction of structurally stable systems; in its turn, the notion of structural stability had been introduced much earlier by Andronov and Pontryagin [**AP**,1937]. As conjectured in the late 1960's and only recently proved, it turns out that the two notions are *essentially equivalent* to each other, at least for C^1 diffeomorphisms of a compact manifold. *Of course, for stability one also has to impose the transversality of all stable and unstable manifolds or, for limit-set stability, the no-cycle condition; see concepts below.* It is, however, the hyperbolicity of the limit set which is the main ingredient in this comparison. The solution of this well known conjecture and other related results that we state here go beyond what is needed to understand the next chapters of this book. It is, however, enlightening, in the study of bifurcations (meaning loss of stability), to be acquainted with the fact that the notions of stability and hyperbolicity are that much interconnected.

The concept of (structural) stability deals with the topological persistence of the orbit structure of a dynamical system (endomorphism, diffeomorphism or flow) under small perturbations. (Notice the difference with the concept of Lyapunov stability which concerns attracting sets of a given system). The persistence of the orbit structure is expressed in terms of a *homeomorphism* of the ambient manifold sending orbits of the initial system onto orbits of the perturbed one. If this can be done for any C^k−small ($k \geq 1$) perturbation, then we call the system C^k (*structurally*) *stable* or *globally stable*. Here we are mostly concerned with diffeomorphisms, in which case we require this orbit preserving homeomorphism to be a *conjugacy*. That is, if φ is the initial map, $\tilde{\varphi}$ a C^k−small perturbation of it, we then require the existence of a homeomorphism h such that $h\varphi(x) = \tilde{\varphi}h(x)$ for all x in the ambient manifold. We do not require the conjugacy to be differentiable, for otherwise we would impose invariance of eigenvalues of the linear part of the diffeomorphism at fixed (periodic) points. The same concept applies to endomorphisms, i.e. maps from the ambient manifold to itself.

We will be treating here the case of n-dimensional diffeomorphisms; for diffeomorphisms of the circle (and flows on surfaces) there are early important works of Pliss [**P**,1960], Arnold [**A**,1961a] and specially Peixoto [**P**,1962]. Often, we are concerned with stability restricted to the main part of the orbit structure, the limit set or nonwandering set. Let us recall these concepts.

Let $\varphi \colon M \to M$ be a C^k ($k \geq 1$) diffeomorphism of a compact, boundaryless, smooth manifold of arbitrary dimension. For $x \in M$, we define the α and ω-limit sets as

$$\alpha(x) = \{y \in M \mid \exists n_i \to -\infty \text{ such that } \varphi^{n_i}(x) \to y\},$$
$$\omega(x) = \{y \in M \mid \exists n_i \to +\infty \text{ such that } \varphi^{n_i}(x) \to y\}.$$

The *positive* and *negative limit sets* are then defined as $L^+(\varphi) = \overline{\bigcup_{x \in M} \omega(x)}$ and $L^-(\varphi) = \overline{\bigcup_{x \in M} \alpha(x)}$; the *limit set* $L(\varphi)$ is the union of $L^+(\varphi)$ and $L^-(\varphi)$. From the definitions, it is clear that $L^+(\varphi)$ and $L^-(\varphi)$ are φ-invariant, i.e. $\varphi(L^+(\varphi)) = L^+(\varphi)$ and $\varphi(L^-(\varphi)) = L^-(\varphi)$. Moreover, for each $x \in M$, $\varphi^n(x)$ approaches $L^+(\varphi)$ or $L^-(\varphi)$ as $n \to \infty$ or $n \to -\infty$. So $L^+(\varphi)$ and $\varphi|L^+(\varphi)$, $L^-(\varphi)$ and $\varphi|L^-(\varphi)$, describe the asymptotic behaviour of orbits, i.e. sequences $\{\varphi^n(x)\}$, in M for $n \to \infty$ or $n \to -\infty$. Another relevant concept is that of *nonwandering* point: x is nonwandering if for any neighbourhood U of it, there is an integer n such that $\varphi^n(U) \cap U \neq \phi$. Again, the union of the nonwandering points, which is denoted by $\Omega(\varphi)$, is a φ-invariant compact set. Clearly, all α or ω-limit points as well as homoclinic points (see Chapter 1) are nonwandering. In Section 4, Chapter 5, we provide an example of a homoclinic tangency which is in $L^+(\varphi)$ but not in $L^-(\varphi)$, or vice versa; this example also shows that, in general, the nonwandering and limit sets are different. However, as we shall see in Chapter 2, any transversal homoclinic point is an accumulation point of periodic orbits and so it is in $L^+(\varphi)$, in $L^-(\varphi)$, and in $\Omega(\varphi)$. Finally, we define another useful concept: the chain recurrent set, which is the union of the chain recurrent points. A point p is chain recurrent if for each $\varepsilon > 0$ there are points $x_0 = p, x_1, x_2, \cdots, x_k = p$ such that $d(f(x_{i-1}), x_i) < \varepsilon$ for $1 \leq i \leq k$, d being a distance function.

If $L^+(\varphi)$ (or $L^-(\varphi)$) is hyperbolic (see Chapter 2 and Appendix 1), then one can show that $\overline{\text{Per}\,(\varphi)} = L^+(\varphi)$ (or $L^-(\varphi)$), where Per (φ) indicates the set of periodic points of φ; one can then write as in [**N**,1972]:

$$L^+(\varphi) = \Lambda_1 \cup \cdots \cup \Lambda_k$$

where each Λ_i is *invariant*, *compact*, *transitive* (it has a dense orbit) and has a *dense subset of periodic orbits*. This is called the spectral decomposition of $L^+(\varphi)$. Moreover, by [**HPPS**, 1970] (see also [**N**,1980], [**B**,1977] for a different and relevant proof using the idea of "shadowing" of orbits), each Λ_i is the *maximal invariant set* in a neighbourhood of it. This last fact is actually equivalent to what we call *local product structure* in Λ_i: there exist $\varepsilon > 0$ and $\delta > 0$ such that if the distance between $x, y \in \Lambda_i$ is smaller than δ then their local stable and unstable manifolds of size ε (see Appendix 1) intersect each other in a unique point and this point is in Λ_i. Also, one can prove that if $\omega(x) \subset \Lambda_i$ then $x \in W^s(z)$ for some $z \in \Lambda_i$. In general, a set with the properties above is called a *basic set* for the diffeomorphism.

If we assume that the nonwandering set $\Omega(\varphi)$ is hyperbolic and $\overline{\text{Per}\,(\varphi)} = \Omega(\varphi)$, then we say that φ satisfies *Axiom A*. In this case we have $\Omega(\varphi) = L^+(\varphi)$ and so we can write the nonwandering set as a finite union of basic sets. This is the content of Smale's spectral decomposition theorem [**S**,1970]; the corresponding version for the limit set as presented above appeared later in [**N**,1972]. Notice that if $\Lambda_1 \cup \cdots \cup \Lambda_k$ is the spectral decomposition of $L^+(\varphi)$ (or $\Omega(\varphi)$) then $M = \bigcup_i W^s(\Lambda_i)$, where $W^s(\Lambda_i) = \{y \,|\, \omega(y) \subset \Lambda_i\}$ is called the stable set of Λ_i; as discussed above $W^s(\Lambda_i) = \bigcup_{x \in \Lambda_i} W^s(x)$. Similar statements are valid for the unstable sets of the Λ_i's, corresponding to a spectral decomposition of $L^-(\varphi)$ or $\Omega(\varphi)$. Some $W^s(\Lambda_i)$ must be open; in this case Λ_i is called an *attractor*. (A more general definition of attractor is in the next section). Dually if $W^u(\Lambda_i)$ is open, then we say that Λ_i is a *repeller*. Finally, Λ_i is of *saddle type* if it is neither an attractor nor a repeller. Another property of Axiom A diffeomorphisms: the stable sets of attractors cover an open and dense subset of M and the same is true for unstable sets of repellers. It is an interesting fact that if φ is C^2, then the union of the stable sets of attractors has total Lebesgue measure; see Ruelle [**R**,1976] and Bowen–Ruelle [**BR**,1975]. There are, however, examples of C^1 saddle-type basic sets with stable sets of positive Lebesgue measure [B,1975b], which are detailed in Chapter 4. Another interesting fact about basic sets is that they are *expansive*: for each basic set Λ there is a constant $\alpha > 0$ such that for each pair of different points in Λ, their (full) orbits get apart by at least α. *From this it follows that hyperbolic attractors which are not just fixed or periodic sinks have sensitive dependence on initial conditions*: for most pairs of different points in the stable set of such an attractor Λ, the positive orbits get apart by at least a constant (which depends on the attractor). *Most* here means probability 1 in $W^s(\Lambda) \times W^s(\Lambda)$.

The following relevant result concerning basic sets states that they are persistent under C^k–small perturbations (see Appendix 1); in particular, hyperbolic attractors are persistent.

THEOREM 1. *If Λ is a basic set for a C^k diffeomorphism $\varphi\colon M \to M$, then for any $\tilde\varphi$ close to φ its maximal invariant set in some neighbourhood of Λ is a basic set $\tilde\Lambda$ and $\varphi|\Lambda$ is conjugate to $\tilde\varphi|\tilde\Lambda$. Moreover, if we require the conjugacy to be C^0–close to the inclusion map of Λ into M, then it is unique and it is in fact Hölder continuous.*

Usually we call this set $\tilde\Lambda$ the "smooth" or "analytic" continuation of Λ for a given perturbation of φ. The result can be applied when all of M is hyperbolic for φ; in this case we say that φ is Anosov. As a corollary, we have the following

THEOREM 2 [A,1967]. *Anosov diffeomorphisms are globally C^k-stable.*

Moser's elegant proof of this last result [M,1969] actually suggested the original proof of the first theorem, but of course Anosov's theorem was shown before. Nowadays there is a simple way of showing the existence of the conjugacy: one uses again the idea of shadowing of orbits mentioned above.

Another relevant class of systems in our context is that of Morse–Smale diffeomorphisms. We call φ Morse-Smale if

(i) $\Omega(\varphi)$ consists of a finite number of periodic orbits, all of them hyperbolic,

(ii) φ satisfies the transversality condition: the stable and unstable manifolds of the elements in $\Omega(\varphi)$ are all in general position.

It turns out that in (i) above one can write $L^+(\varphi)$ or $L^-(\varphi)$ instead of $\Omega(\varphi)$. *Morse–Smale diffeomorphisms are abundant in the sense that they contain the time-1 maps of an open and dense subset of gradient vector fields on every manifold.*

THEOREM 3 [PS,1970]. *Morse–Smale diffeomorphisms are C^k-stable. (In particular, there are stable diffeomorphisms on every manifold.) Conversely,* among diffeomorphisms whose nonwandering sets consist of finitely many periodic orbits, hyperbolicity of these orbits and transversality of their stable and unstable manifolds are necessary for C^k-stability.

In view of these results, it seemed that hyperbolicity of the nonwandering set (or limit set) and transversality of stable and unstable manifolds were the precise conditions that should grant C^k-stability of the diffeomorphism. So, let us say that an Axiom A diffeomorphism f satisfies the transversality condition if the stable and unstable manifolds of any two points in $\Omega(f)$ are in general position.

STABILITY CONJECTURE [**PS**,1970]: A C^k (or C^s, $s \geq k$) diffeomorphism is C^k-stable if and only if it satisfies Axiom A and the transversality condition.

REMARK 1: We can phrase the stability conjecture in terms of the limit set (or just the positive or negative limit set): a diffeomorphism is C^k-stable if and only if its limit set is hyperbolic and the transversality condition holds. This equivalent, and perhaps more elegant, statement is further commented on below.

Much in parallel, let us start discussing stability restricted to the nonwandering set or to the limit set. We say that a diffeomorphism φ is $C^k-\Omega$-stable if there exists a C^k neighbourhood of it such that if $\tilde{\varphi}$ belongs to this neighbourhood, then $\varphi|\Omega(\varphi)$ is conjugate to $\tilde{\varphi}|\Omega(\tilde{\varphi})$. Similarly for the limit set. Let now φ be an Axiom A diffeomorphism and let $\Omega(\varphi) = \Lambda_1 \cup \cdots \cup \Lambda_k$ be its spectral decomposition. A j-cycle on Ω is a string of j pairs of points $x_1, y_1 \in \Lambda_{i_1}, \cdots, x_j, y_j \in \Lambda_{i_j}$, with not all $i_1, \cdots, i_j$ equal, such that $W^u(y_1) \cap W^s(x_2) \neq \phi, \cdots, W^u(y_j) \cap W^s(x_1) \neq \phi$. If the limit set, or just the positive or negative limit set, is hyperbolic then we also have a spectral decomposition for it and the notion of cycles can be applied. In either case, when there are *no cycles* we can construct a *filtration* ([**S**,1967],[**C**,1978]): a sequence $M_0 = \phi$, $M_1 \subset M_2 \subset \cdots \subset M_k = M$ of compact submanifolds with boundary for $0 < i < k$ such that $\varphi(M_i) \subset \text{Int } M_i$ and in $M_{i+1} - M_i$ the maximal invariant set is Λ_i. The existence of filtrations implies the following proposition, which clarifies why in the statement of stability results and conjectures we can mention either Axiom A (hyperbolicity of the nonwandering set and density of periodic orbits) or just hyperbolicity of the limit set or the chain recurrent set; see [**N**,1972; **FS**,1977].

PROPOSITION 1. *If $L(\varphi)$ (or just $L^+(\varphi)$ or $L^-(\varphi)$) is hyperbolic and there are no cycles on it, then $\Omega(\varphi) = L(\varphi)$ and so φ satisfies Axiom A (and there are no cycles on $\Omega(\varphi)$). If the chain recurrent set of φ is hyperbolic then φ satisfies Axiom A and the no-cycle condition.*

The existence of a filtration implies a global control on the nonwandering or limit set when we perturb the map: there appear no nonwandering points far from the original ones. This fact and the persistence of basic sets stated above imply Smale's Ω-stability theorem [S,1970].

THEOREM 4. *The set of diffeomorphisms satisfying Axiom A and the no-cycle property is open in $Diff^k(M)$ and its elements are Ω-stable.*

In the way of a converse to this theorem, we have the following

THEOREM 5 [P,1970]. *If φ is an Axiom A diffeomorphism and there is a cycle on $\Omega(\varphi)$, then φ is not Ω-stable. In fact, there are arbitrarily close diffeomorphisms $\tilde{\varphi}$ such that Per $(\varphi) \not\subseteq$ Per $(\tilde{\varphi})$. A similar statement is true for diffeomorphism with hyperbolic limit sets.*

COROLLARY 1. *If φ and all nearby diffeomorphisms have their nonwandering sets (limit sets) hyperbolic, then they are Ω-stable. On the other hand, if φ or arbitrarily close diffeomorphisms exhibit homoclinic tangencies (see Chapter 2 or 3) then φ is not Ω-stable.*

In view of the results above we formulate the following conjecture.

Ω-STABILITY CONJECTURE: A diffeomorphism φ is $C^k - \Omega$-stable if and only if it satisfies Axiom A and the no-cycle property.

Back to the stability conjecture, in the early 1970's Robbin [R,1971] proved that diffeomorphisms satisfying Axiom A and the transversality condition are C^k-stable for $k \geq 2$. Soon afterwards, de Melo [M,1973b] proved the result for C^1 diffeomorphisms of surfaces using ideas close to [PS, 1970]. By 1976, Robinson [R,1976] completed the result for $k = 1$ in any dimension using an approach somewhat different from the previous ones. Before, in [R,1974], he had the corresponding version for flows with $k \geq 2$. In another paper [R,1973], he also proved that the transversality condition is necessary for C^k stability.

So, in both the stability and the Ω-stability conjectures it remained to show that hyperbolicity of the nonwandering set was necessary for either kind of stability. This was the missing fact to establish such a fundamental link between stability or Ω-stability and hyperbolicity of the nonwandering (or limit) set. From the beginning it was clear that with the available knowledge this goal was probably beyond reach for C^k stability or Ω-stability when $k \geq 2$. In fact, it is still unknown whether $\overline{\text{Per }(\varphi)} = \Omega(\varphi)$ if φ is $C^2 - \Omega$-stable; for $k = 1$, this follows from Pugh's closing lemma [P,1967]. By 1980, Mañé concluded both conjectures for C^1 surface diffeomorphisms [M,1982]; independently, Liao [L,1980] and Sannami [S,1983] obtained this same result. For flows, Liao seems to have made substantial progress toward the same conclusion on 3-manifolds; in higher dimensions the question is still open. It is interesting to observe that the situation looks rather different for manifolds with boundary and flows leaving the boundary invariant: there exist singular horseshoes that are stable but not hyperbolic [LP,1986].

Finally, just about 20 years after it was proposed, Mañé in a remarkable paper [**M**,1988] presented a solution of the C^1 stability conjecture for diffeomorphisms in any dimension. His proof, however, did not include the $C^1 - \Omega$-stability conjecture because he needed the transversality condition (typical of stability but not of Ω-stability) to complete his arguments. This was done in [**P**,1988], arguing instead just with the no-cycle property.

Actually, one may ask if the following set of equivalences is true:

$f \in \mathcal{D}^1(M) \Longleftrightarrow f$ satisfies Axiom A and the no-cycle property $\Longleftrightarrow f$ is Ω-stable.

Here, $\mathcal{D}^1(M)$ denotes the interior, with respect to the C^1 topology, of the set of diffeomorphisms whose periodic orbits are all hyperbolic. From the results above it only remained to show that if f is in $\mathcal{D}^1(M)$ then it satisfies Axiom A. The truth of this statement, and thus of the set of equivalences above, was recently and independently announced by Aoki [**A**,1991] and Hayashi [**H**,1991].

Closing this section, we want to pose two questions that are somewhat inspired by the results above and are relevant in the context of homoclinic bifurcations, the main topic of this text. They concern differentiable arcs or one-parameter families φ_μ of C^k diffeomorphisms such that φ_μ satisfies Axiom A and the transversality condition for $\mu < \mu_0$ and φ_{μ_0} is Ω-unstable; such μ_0 is called the *first bifurcation point* of φ_μ.

The first question is: what types of bifurcation can occur for φ_{μ_0} if the family φ_μ is *generic*, i.e. if the family φ_μ belongs to some *residual (Baire second category)* class of families ? We *conjecture* that in *two dimensions*, we only have three possibilities: φ_{μ_0} has a nonhyperbolic periodic orbit, a homoclinic tangency, or a heteroclinic tangency involving periodic points in a cycle. *In higher dimensions* there are more cases, like *homoclinic tangencies of basic sets:* $W^s(x)$ tangent to $W^u(y)$ for x, y in the same basic set, neither point being necessarily periodic. A main difference is that the *boundary* of a basic set in two dimensions is made of stable and unstable manifolds of periodic orbits; see Appendix 2.

The next question concerns a generic family φ_μ on a surface such that φ_μ is Morse–Smale for $\mu < \mu_0$ and there exist values $\tilde{\mu} > \mu_0$ arbitrarily near μ_0 such that $\varphi_{\tilde{\mu}}$ has infinitely many periodic orbits. The problem now is whether there is $\mu_1 > \mu_0$ near μ_0 so that φ_{μ_1} exhibits some homoclinic tangency associated to some periodic point (see Chapter 3). In higher dimensions we formulate the question in terms of homoclinic bifurcations as in Chapter 7: one may persistently (open set of families) create homoclinic orbits without creating homoclinic tangencies at all!

We refer the reader to [**NP**,1976; **NP**,1973] where similar questions were studied and a version of the above conjecture posed.

§2   Sensitive chaotic dynamics

The notion of "*chaos*" in dynamical systems, as opposed to theology and the usual meaning of total disorder, refers to a situation where (forward) orbits do not converge to a periodic or quasi-periodic orbit and where *the evolution of the orbits has some degree of unpredictability* or *their behaviour is sensitive with respect to initial conditions*. Although this phenomenon was theoretically known, in particular for nontrivial hyperbolic attractors, it came to many as a surprise that it also appeared in numerically generated orbits of quite simple systems. Among the first such examples, which were investigated numerically, there were the Lorenz attractor [**L**,1963], the logistic map [**CE**,1980] and the Hénon map [**H**,1976]–in fact all these systems depend on parameters and for a substantial set of parameter values these sensitive or chaotic phenomena appear. All these examples are nonhyperbolic and a big effort was needed to get some theoretical understanding of them. The last two examples play an important role in the dynamics at homoclinic bifurcations; see Chapters 3, 6 and 7. In this section we shall formalize the main notions involved in "chaos" and indicate instances where they occur in hyperbolic systems. We want to point out that there are different formalizations of these notions (although there is agreement about the general flavour).

Here, we restrict ourselves to dynamical systems defined by a continuous map $\varphi\colon M \to M$, where we assume M to be a compact metric space—if M is not compact the discussion still makes sense if we restrict to points in M whose positive orbits have a compact closure and, hence, are bounded. We say that the (positive) *orbit* $\{x, \varphi(x), \varphi^2(x), \cdots\}$ of x is *sensitive* or *chaotic* if there is a positive constant $C > 0$ such that for any $q \in \omega(x)$ (for the definition see the previous section) and any $\varepsilon > 0$ there are integers $n_1, n_2, n > 0$ such that $d(\varphi^{n_1}(x), q) < \varepsilon$, $d(\varphi^{n_2}(x), q) < \varepsilon$, but $d(\varphi^{n_1+n}(x), \varphi^{n_2+n}(x)) > C$. We observe that an orbit asymptotic to a (quasi-) periodic one is not chaotic in the above sense: for such orbits, if $\varphi^{n_1}(x)$ and $\varphi^{n_2}(x)$ are close, then $\varphi^{n_1+n}(x)$ and $\varphi^{n_2+n}(x)$ remain close for all $n > 0$. Also, a sensitive orbit in the above sense is unpredictable to the extent that if we know that some y on the positive orbit of x is extremely close to $q \in \omega(x)$, this is not enough to predict, say within distance C, all future iterates of y. *This last fact is closely related with the sensitive dependence on initial conditions discussed in the previous section.* Similarly to the fact that in the stable set of a *nontrivial hyperbolic attractor* one has sensitive dependence on initial conditions, one can prove that most points in such a stable set have *sensitive orbits*–the set of points with chaotic orbits even has *total Lebesgue measure* in the stable set; see [**ER**, 1985]. Not only in hyperbolic attractors are there sensitive orbits, but also in nontrivial hyperbolic sets of saddle-type: this can easily be seen using symbolic dynamics as introduced for the horseshoe example in Chapter 2. The difference is that in this last case the set of points with sensitive orbits has Lebesgue measure zero.

For the case where M is a manifold, and hence the notion "Lebesgue measure zero" is defined, we say that the *dynamical system* defined by φ is *sensitive*, or that φ has *sensitive* or *chaotic dynamics*, if the set of points with sensitive or chaotic orbits has *positive Lebesgue measure*.

Up to now, the notion of sensitive or chaotic dynamics has mainly been defined negatively (?!): namely, in terms of unpredictability. From the numerical experiments, mentioned before, it was however apparent that certain aspects of the positive orbits were very predictable: for any initial point (in some open set), one always finds numerically the *same* ω-limit set. Thus, chaos or sensitivity as observed above should not be interpreted as *total unpredictability*: instead, it should mean that most orbits of the system will finally be getting apart from each other and continue to wander around a "larger" but definite set. This leads to the definition of a *strange attractor*. We say that a compact set $A \subset M$ is a *strange attractor* if there is an open set U with a subset $N \subset U$ of Lebesgue measure zero, such that for all $x \in U \setminus N$, $\omega(x) = A$ and the (positive) orbit of x is chaotic. (We allow for the exceptional set N because even for hyperbolic attractors a dense set of points, of Lebesgue measure zero, is attracted to periodic orbits; and *as long as N has Lebesgue measure zero it should not be of influence on numerical experiments*). Another and even more common definition of *strange attractor* is to require A to have sensitive dependence on initial conditions on U; see the discussions in Chapter 7.

In many cases, including the numerical examples, one is interested in the persistence of phenomena not only under a (small) perturbation of the initial point (of a positive orbit) but also under a small perturbation of the map $\varphi: M \to M$. Intuitively, one says that the dynamics of φ is *persistently sensitive or chaotic* if small perturbations of φ have, with positive probability, sensitive dynamics. But the problem with this intuitive notion is that there is no "natural" measure on the set of maps $\varphi: M \to M$. On the other hand, if we would require all small perturbations of φ to have sensitive dynamics (which is the case for a nontrivial hyperbolic attractor and whose dynamics is called for this reason *fully persistently sensitive or chaotic*) we would exclude important cases like the *logistic map* for many values of the parameter. There is, however, one important instance in which this notion can be formally defined: if we are in a context where the notion of generic k-parameter unfoldings $\varphi_{\mu_1,\dots,\mu_k}$ of φ is defined (with $\varphi_{0,\dots,0} = \varphi$), we say that φ has persistently sensitive dynamics if for any such generic k-parameter unfolding, the set of $\mu = (\mu_1, \dots, \mu_k)$ values, for which φ_μ has sensitive dynamics, has positive Lebesgue measure.

In the last section of Chapter 6 and in Chapter 7 we shall discuss the consequences of our investigations of homoclinic bifurcations in terms of the above notions of sensitive or chaotic dynamics.

CHAPTER 1

EXAMPLES OF HOMOCLINIC ORBITS
IN DYNAMICAL SYSTEMS

We discuss a number of dynamical systems with (transverse) homoclinic orbits, just to have some examples in order to motivate the following chapters. First we need some definitions. We deal with diffeomorphisms $\varphi\colon M \to M$ of a compact manifold to itself. In this chapter it is enough to assume that φ is of class C^1, but for some of the later results we need φ to be C^2 or C^3. Also the compactness of M is not always needed—some of the examples in this chapter will be on $\mathbb{R}^2$.

We say that $p \in M$ is a hyperbolic fixed point of φ if $\varphi(p) = p$ and if $(d\varphi)_p$ has no eigenvalue of norm 1. For such a hyperbolic fixed point, one defines the stable and the unstable manifold as

$$W^s(p) = \{x \in M | \varphi^i(x) \to p \text{ for } i \to +\infty\}$$

and

$$W^u(p) = \{x \in M | \varphi^i(x) \to p \text{ for } i \to -\infty\}.$$

According to the invariant manifold theorem (see [**HPS**,1977] and also Appendix 1) both $W^s(p)$ and $W^u(p)$ are injectively immersed submanifolds of M, are as differentiable as φ, and have dimensions equal to the number of eigenvalues of $(d\varphi)_p$ with norm smaller than 1, respectively bigger than 1. One can give the corresponding definitions for periodic points, i.e. fixed points of some power of φ.

If $\varphi\colon \mathbb{R}^n \to \mathbb{R}^n$ is a linear map with no eigenvalues of norm 1, then the origin 0 is a hyperbolic fixed point and $W^s(0)$, $W^u(0)$ are complementary linear subspaces: $\mathbb{R}^n = W^s(0) \oplus W^u(0)$.

We say that if p is a hyperbolic fixed point of φ, q is *homoclinic* to p if $p \neq q \in W^s(p) \cap W^u(p)$, i.e. if $q \neq p$ and if $\lim_{i\to\pm\infty} \varphi^i(q) = p$ (this last form of the definition makes clear why Poincaré called such points "*bi-asymptotique*"). We say that q is a *transverse homoclinic* point if $W^s(p)$ and $W^u(p)$ intersect transversally at q, i.e. if

$$T_q(M) = T_q(W^s(p)) \oplus T_q(W^u(p)).$$

It is clear that *linear diffeomorphisms* have *no homoclinic points*.

§1   Homoclinic orbits in a deformed linear map

We start with the linear map $\varphi\colon \mathbb{R}^2 \to \mathbb{R}^2$, $\varphi(x,y) = (2x, \frac{1}{2}y)$. The stable manifold is the y-axis, the unstable manifold is the x-axis. Next consider the composition $\Psi \circ \varphi$, where $\Psi\colon \mathbb{R}^2 \to \mathbb{R}^2$ is a diffeomorphism of the form

$$\Psi(x,y) = (x - f(x + y), y + f(x + y)),$$

where f is some smooth function. This means that Ψ is pushing points along lines of the form $\{x + y = c\} = \ell_c$ over a distance which only depends on c. We take f a smooth function which is zero on $(-\infty, 1]$ and such that $f(2) > 2$. In this case the stable and unstable manifolds $W^s(0)$ and $W^u(0)$ for the diffeomorphism $\Psi \circ \varphi$ intersect each other outside the origin.

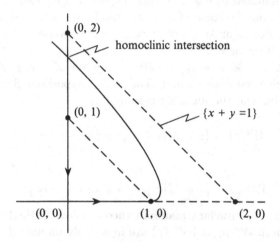

Figure 1.1

In fact, due to the construction, $\{(x,y)|x = 0, y \le 2\}$ belongs to $W^s(0)$ and $\{(x,y)|x \le 1, y = 0\}$ belongs to $W^u(0)$. Also

$$\Psi(\{(x,y)|1 \le x \le 2, y = 0\})$$

belongs to $W^u(0)$. From this and the description of Ψ we obtain a homoclinic intersection; see Figure 1.1. By choosing f appropriately we can produce a *transverse* homoclinic point. Not much can be said at this point about the global configuration of $W^s(0)$ and $W^u(0)$ but certainly this configuration will be very complicated as shown later in this chapter.

§2   The pendulum

Our next example, the *pendulum*, contains a line of non transverse homoclinic points. Consider the differential equation

$$\ddot{\theta} = -\sin\theta, \quad \theta \in \mathbb{R}/2\pi,$$

which defines a system of ordinary differential equations on the annulus:

$$\dot\theta = y,$$
$$\dot y = -\sin\theta, \qquad (1)$$

with $\theta \in \mathbb{R}/2\pi$, $y \in \mathbb{R}$.

We take the time T map of this system, i.e. the diffeomorphism φ such that $\varphi(\theta, y) = (\tilde\theta, \tilde y)$ whenever there is a solution $(\theta(t), y(t))$ of (1) with $(\theta(0), y(0)) = (\theta, y)$ and $(\theta(T), y(T)) = (\tilde\theta, \tilde y)$. Then the fixed points of φ are $(\theta = 0, y = 0)$ and $(\theta = \pi, y = 0)$. The first is not hyperbolic (the eigenvalues of $(d\varphi)_{(0,0)}$ have norm 1) but the second is: it has a one-dimensional stable and a one-dimensional unstable manifold. In order to determine the positions of these stable and unstable manifolds it is important to note that the function

$$E(\theta, y) = -\cos\theta + \frac{1}{2}y^2$$

is constant along solutions of (1): it is the energy, $-\cos\theta$ being the potential energy and $\frac{1}{2}y^2 = \frac{1}{2}\dot\theta^2$ being the kinetic energy. This means that both $W^u(\pi, 0)$ and $W^s(\pi, 0)$ are given by

$$-\cos\theta + \frac{1}{2}\, y^2 = 1$$

see Figure 1.2.

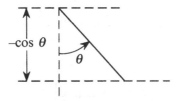

Figure 1.2

In Figure 1.3 this homoclinic line is indicated together with some other energy levels.

By a small perturbation of φ one can make $W^u(\pi, 0)$ and $W^s(\pi, 0)$ intersect transversally (using a perturbation as in the first example, or referring to the proof of the Kupka–Smale theorem [**S**,1963; **K**,1964]). Such perturbations also arise as a consequence of external forcing, e.g. through replacing (1) by

$$\dot\theta = y,$$
$$\dot y = -\sin\theta + \varepsilon \cdot \cos(2\pi t/T).$$

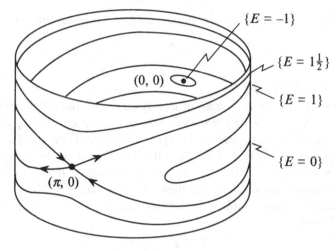

Figure 1.3

§3   The horseshoe

In the following example, the *horseshoe* (see Smale [**S**,1965]), we have transverse homoclinic points and still we are fairly well able to describe globally the stable and the unstable manifold. In order to describe the diffeomorphism, let Q be a square in $\mathbb{R}^2$ and let φ map Q as indicated in Figure 1.4, such that on both components of $Q \cap \varphi^{-1}(Q)$, φ is affine and preserves both horizontal and vertical directions, and such that 1,2,3 and 4 are mapped to $1'$, $2'$, $3'$ and $4'$.

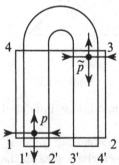

Figure 1.4

In Q, φ has two fixed points p and $\tilde{p}$ as indicated; we restrict our attention to p. Since φ is affine on $Q \cap \varphi^{-1}(Q)$, the stable and unstable manifolds $W^s(p)$ and $W^u(p)$, near p, are straight lines. In order to find the continuation one has to iterate φ^{-1} (for $W^s(p)$) and φ (for $W^u(p)$). Inside Q this gives pieces of straight lines, horizontal for $W^s(p)$ and vertical for $W^u(p)$. With a few iterations one gets Figure 1.5.

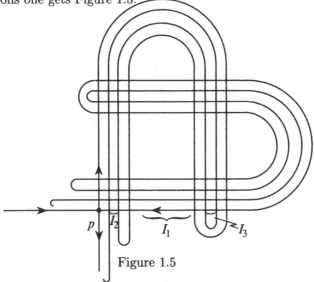

Figure 1.5

As a final remark on this example, observe that however far we iterate, the intervals $I_1, I_2, I_3 \subset W^s(p)$ will never be intersected by $W^u(p)$. There is in fact a countably infinite number of such open intervals, and $W^s(p) \cap W^u(p)$ consists of the boundary points of a Cantor set in $W^s(p)$ which is the complement of these intervals.

§4 A homoclinic bifurcation

We speak of a homoclinic bifurcation if in a one-parameter family of diffeomorphisms, a pair of homoclinic intersections collides, forms a tangency and then disappears, or, reversing the direction, if a pair of homoclinic points is generated after a tangency.

Such bifurcations are obtained from the previous example by composing φ with a map $(x, y) \mapsto (x, y - \mu)$, which slides the image, in particular the image of Q, down. In Figure 1.6 we show the effect of this sliding on the geometry of the stable and unstable manifold $W^s(p_\mu)$ and $W^u(p_\mu)$ for increasing values of μ. In Figure 1.6(a) we have just the previous example.

In Figure 1.6(b) one sees the first nontransverse homoclinic orbit (four iterations are indicated). From Figure 1.6(c), it is clear that near one such homoclinic bifurcation there are many others—see also Chapter 3.

§5 Concluding remarks

The complexity of the configuration of stable and unstable manifolds in the examples is typical for the case where one has at least one transverse

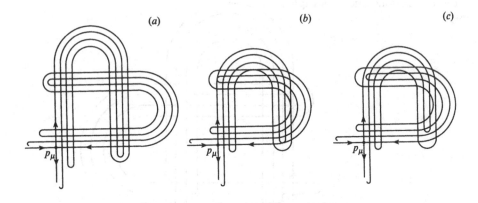

Figure 1.6

homoclinic point. Paraphrasing Poincaré, one can convince oneself of this complexity by trying to draw examples, keeping in mind that

- $W^s(p)$ and $W^u(p)$ are φ-invariant, i.e. $\varphi(W^s(p)) = W^s(p)$ and $\varphi(W^u(p)) = W^u(p)$,

- $W^s(p)$ and $W^u(p)$ have no self intersections,

- near p, φ is well approximated by the linear map $(d\varphi)_p$, which leads to the following consequence (λ−lemma, [**P**,1969]; see also Appendix 1): if ℓ is a smooth curve intersecting $W^s(p)$ transversally then the forward images $\ell_i = \varphi^i(\ell)$ contain compact arcs $m_i \subset \ell_i$ which approach differentiably a compact arc m in $W^u(p)$, as illustrated in Figure 1.7.

For higher- or even infinite-dimensional diffeomorphisms the situation is basically the same.

The first time that transverse homoclinic points were constructed was by Poincaré in his prize essay [**P**,1890]. The existence of these homoclinic points implied the non-convergence of certain power series expressions for solutions of a Hamiltonian system, which was comparable with the Hamiltonian sytem describing the restricted three-body problem. This indicated that certain qualitative information, like "stability", was not obtainable by these analytic power series methods.

Later it was realized that the dynamics of a diffeomorphism φ, or the topology of its orbits, shows a great complexity if and only if φ has some hyperbolic periodic point with a homoclinic intersection of its stable and unstable manifolds. The same can be said about *chaotic orbits*; see Chapter 0, Section 2: they occur if and only if there is some homoclinic orbit (but a homoclinic orbit does not need to imply *chaotic dynamics*). *Although*

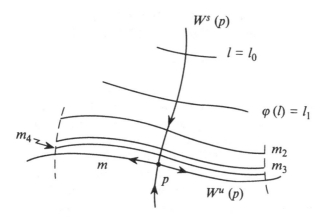

Figure 1.7

these last statements are in no way theorems (and can be expected to be true only in a generic sense), many facts discussed in these notes can actually be interpreted as partial results in this direction, see Chapter 7. On the other hand, we also discuss "stability results" concerning the dynamics in the presence of transverse homoclinic points. They deal with the not so infrequent situation that the dynamics of a diffeomorphism φ (with transverse homoclinic points), although very complicated, remains unchanged in a topological sense when φ is slightly perturbed in the C^1 topology; see Chapter 0.

CHAPTER 2

DYNAMICAL CONSEQUENCES OF A TRANSVERSE
HOMOCLINIC INTERSECTION

In this chapter we analyse the dynamical complexity due to one transverse homoclinic orbit. Although our discussion refers to the two-dimensional situation, the results and their proofs can be extended to arbitrary dimensions with minor modifications and even to Banach spaces (or Banach manifolds).

Since one of the main features of a transverse homoclinic intersection consists of the occurrence of hyperbolic invariant sets, this chapter relies to some extent on Appendix 1, and even seems to overlap with the subject matter of that appendix. The emphasis however is different. Here we concentrate on the *geometric* properties of invariant sets near transverse homoclinic orbits which we show to be hyperbolic, and to do this we make use of some results from the *analytical* theory of hyperbolicity in Appendix 1. For a more complete view on hyperbolic sets in dimension 2 one should also consult Appendix 2 on Markov partitions. To present the main ideas in a more transparent way, we will assume the diffeomorphisms to be of class C^2 although the results are still true in the C^1 category: the proofs are similar but the computations more involved in the C^1 case.

§1 Description of the situation – linearizing coordinates and a special domain R

Let $\varphi: M \to M$ be a C^2 diffeomorphism of a surface M and $p \in M$ be a fixed saddle, i.e. $\varphi(p) = p$ and $(d\varphi)_p$ has two real eigenvalues λ and σ with $0 < |\lambda| < 1 < |\sigma|$. For simplicity we assume that $0 < \lambda < 1 < \sigma$. From the theory of hyperbolicity (see Appendix 1) we know that:

- the stable and unstable separatrices of p, $W^s(p)$ and $W^u(p)$, are C^2,

- there are C^1 linearizing coordinates in a neighbourhood U of p, i.e. C^1 coordinates x_1, x_2 such that $p = (0, 0)$ and such that $\varphi(x_1, x_2) = (\lambda \cdot x_1, \sigma \cdot x_2)$. See Figure 2.1.

This linearization follows at once from the existence of φ-invariant stable and unstable foliations near p which are of class C^1; see also [H,1964].

We assume that $W^s(p)$ and $W^u(p)$ have points of transverse intersection different from p—such points, or their orbits, are called *homoclinic* or *bi-asymptotic* to p. In the two-dimensional situation we consider mainly *primary* homoclinic points in order to simplify the figures and the geometric

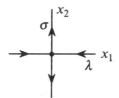

Figure 2.1

arguments. A homoclinic point q is primary if the arcs ℓ^u, joining p and q in W^u, and ℓ^s, joining p and q in W^s, form a double point free closed curve.

Note that whenever p has homoclinic points, it has primary homoclinic points—if all intersections of $W^u(p)$ and $W^s(p)$ are transverse, then the number of primary homoclinic orbits is finite. See Figure 2.2. We also note that the notion of primary homoclinic orbit does not extend to dimensions greater than 2. Still the results which we discuss below extend without much difficulty to the nonprimary or n-dimensional case.

(a)

primary homoclinic points

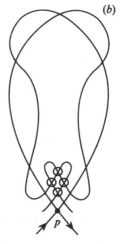

(b)

nonprimary homoclinic points
(encircled)

Figure 2.2

Let the linearizing coordinates be defined on U and let their image be the square $(-1, +1) \times (-1, +1) \subset \mathbb{R}^2$. We consider extensions of the domain of definition of these linearizing coordinates. Identifying points in U with the corresponding points in $\mathbb{R}^2$, we have: if $\varphi^{-1}([\lambda, 1) \times (-1, +1)) \cap U = \phi$, we can extend the domain of the linearizing coordinates x_1, x_2 to $\varphi^{-1}([\lambda, 1) \times (-1, +1))$ using the formulas

$$x_1 = \lambda^{-1} \cdot (x_1 \circ \varphi), \quad x_2 = \sigma^{-1} \cdot (x_2 \circ \varphi).$$

Repeating this construction one can extend the linearizing coordinates along any segment in $W^s(p)$ starting in p: one only has to take the original domain U sufficiently small. This follows from the fact that $W^s(p)$ has no self intersections. In the same way one can extend the domain of these linearizing coordinates along the unstable separatrix $W^u(p)$. Homoclinic intersections, however, form an obstruction to a simultaneous extension of such coordinates along both the stable and the unstable separatrix. In our situation where q is a primary homoclinic point, we extend the linearizing coordinates along both ℓ^u and ℓ^s, the arcs in $W^u(p)$ and $W^s(p)$ joining p and q; however, these coordinates will be bi-valued near q as indicated in Figure 2.3.

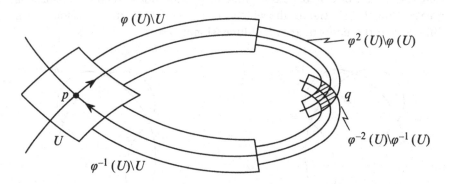

shaded area : bi-valued linearizing coordinates

Figure 2.3

Figure 2.4 shows the situation in $\mathbb{R}^2$: the shaded area denotes the two sets corresponding to the above neighbourhood of q.

Now we consider in the domain of the extended coordinates a rectangle $R = \{-a \leq x_1 \leq b, -\alpha \leq x_2 \leq \beta\}$, $a, b, \alpha, \beta > 0$, containing ℓ^s, the arc in $W^s(p)$ joining p and q, and such that for some N

- $R \cap \varphi^n(R)$ consists of one rectangle containing p for $0 \leq n < N$,
- $R \cap \varphi^N(R)$ consists of two connected components, one containing q, as indicated in Figure 2.5, i.e.

$$\{x_1 = b, -\alpha \leq x_2 \leq \beta\} \cap \varphi^N(R) = \phi,$$
$$\varphi^N(\{-a \leq x_1 \leq b, x_2 = \beta\}) \cap R = \phi.$$

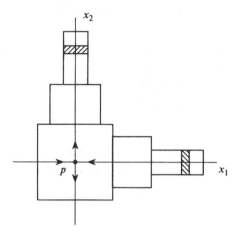

Figure 2.4

For what follows it is important that one can choose R so that N is arbitrarily big: just take β small. By taking β small and hence N big, $\varphi^N(R)$ will become a very narrow strip around ℓ^u. So, transversality of $W^u(p)$ and $W^s(p)$ at q implies transverse intersection of the sides of R and $\varphi^N(R)$.

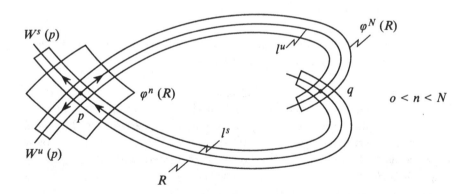

Figure 2.5

The main object of interest in this chapter is the maximal invariant subset of R under φ^N, i.e. the set of those points $r \in R$ such that $\varphi^{k \cdot N}(r) \in R$ for all $k \in \mathbf{Z}$.

§2 The maximal invariant subset of R – topological analysis

From now on we denote φ^N by Ψ. We denote the *maximal invariant subset* of R under Ψ (satisfying the conditions in the previous section) by $\Lambda = \{r \in R | \Psi^k(r) \in R$ for all $k \in \mathbf{Z}\}$. Denoting the corners of R by 1,2,3, and 4 and their images in $\Psi(R)$ by $1', 2', 3'$, and $4'$, the relative positions of R and $\Psi(R)$ are as indicated in Figure 2.6, i.e. the sides of R and $\Psi(R)$ intersect transversally and the topology of the positions of R, its sides and its corners relative to their images under Ψ are as in the figure. We denote the components of $R \cap \Psi(R)$ by $\underline{0}$ and $\underline{1}$, $\underline{0}$ containing p and $\underline{1}$ containing q.

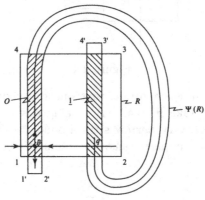

Figure 2.6

THEOREM 1. *For any sequence* $\{a_i\}_{i \in \mathbf{Z}}$, *with* $a_i = 0$ *or 1, there is at least one point* $r \in \Lambda$ *such that* $\Psi^i(r) \in \underline{a_i}$ *for all* $i \in \mathbf{Z}$.

PROOF: We call a closed subset $S \subset R$ a vertical strip if it is bounded (in R) by two disjoint continuous curves ℓ_1 and ℓ_2 connecting the side (1,2) with the side (3,4) (see Figure 2.7). If S is a vertical strip then $\Psi(S) \cap R$ contains two vertical strips, one in $\underline{0}$ and one in $\underline{1}$. Let now $\{a_i\}_{i \in \mathbf{Z}}$ be a sequence as in the theorem. We construct a nested sequence of vertical strips $S_0 \supset S_1 \supset S_2 \supset \cdots$: $S_0 = \underline{a_0}$; S_1 is the vertical strip $\Psi(\underline{a_{-1}}) \cap S_0$; S_2 is the vertical strip $\Psi^2(\underline{a_{-2}}) \cap S_1$; $\dots$; $S_\infty = \bigcap\limits_{i \geq 0} S_i$.

For each point $r \in S_\infty$, $\Psi^{-i}(r) \in \underline{a_{-i}}$, $i \geq 0$.
Horizontal strips are similarly defined and we have horizontal strips $T_1 \supset T_2 \supset T_3 \supset \cdots$ such that for $r \in T_\infty = \bigcap\limits_{i \geq 1} T_i$, $\Psi^i(r) \in \underline{a_i}$ for all $i \geq 1$. Now $S_\infty \cap T_\infty \neq \phi$. Otherwise, for some i_0, $S_{i_0} \cap T_{i_0} = \phi$, but S_{i_0} contains a line from side (1,2) to the side (3,4) and T_{i_0} contains a line from (1,4) to (2,3). These lines have to intersect.

For any point $r \in S_\infty \cap T_\infty$, $\Psi^i(r) \in \underline{a}_i$ for all $i \in \mathbf{Z}$. From this, it also follows that $r \in \Lambda$. □

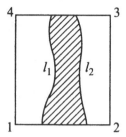

shaded area : vertical strip

Figure 2.7

§3 The maximal invariant subset of R – hyperbolicity and invariant foliations

In this section we impose more conditions on $\Psi = \varphi^N$ restricted to R. In the linearizing coordinates on a neighbourhood of ℓ^s, the arc in $W^s(p)$ joining p and q, we have

$$R = \{-a \leq x_1 \leq b, -\alpha \leq x_2 \leq \beta\};$$

see Section 1. We only have to describe Ψ in those points of R which are mapped back into R, i.e. in $\Psi^{-1}(R) \cap R$. In the component of $\Psi^{-1}(R) \cap R$ containing p and q, Ψ is linear and in fact $\Psi(x_1, x_2) = (\lambda^N x_1, \sigma^N x_2)$ with $0 < \lambda < 1 < \sigma$.

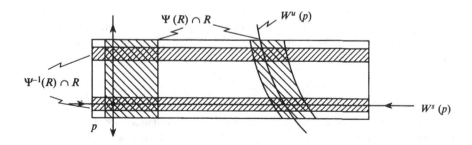

Figure 2.8

The other component of $\Psi^{-1}(R) \cap R$ is mapped to the component of $R \cap \Psi(R)$ containing q. This component of $R \cap \Psi(R)$ is the region where the linearizing coordinates, constructed in Section 1, were bi-valued, or rather where we have apart from the linearizing coordinates following $W^s(p)$, and which are in Figure 2.8 the Cartesian coordinates of the plane, also the linearizing coordinates following $W^u(p)$. We denote by e_1, e_2 the coordinate vector fields of the linearizing coordinates following $W^s(p)$ and by e_1', e_2' the coordinate vector fields of the linearizing coordinates following $W^u(p)$; see Figure 2.9.

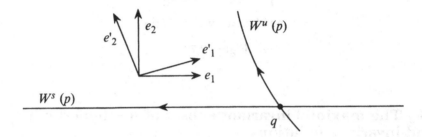

Figure 2.9

For r in the component of $R \cap \Psi^{-1}(R)$ which is mapped on a neigbourhood of q, we have $(d\Psi)e_1(r) = \lambda^N \cdot e_1'(\Psi(r))$ and $(d\Psi)e_2(r) = \sigma^N \cdot e_2'(\Psi(r))$.

Due to the transversality of $W^u(p)$ and $W^s(p)$ and to the thinness of $\Psi(R)$, for N big, e_1 and e_2' are linearly independent. Also, by choosing R and $\Psi(R)$ thin, we may assume that the matrix transforming e_1, e_2 into e_1', e_2' (or its inverse) is almost constant.

THEOREM 1. *For R sufficiently thin, and hence N big, the maximal invariant subset $\Lambda = \bigcap_{n \in \mathbf{Z}} \Psi^n(R)$ in R is hyperbolic. (The technical definition of hyperbolicity is given in Appendix I).*

PROOF: A continuous cone field C on $R \cap \Psi(R)$ is a map which assigns to each $r \in R \cap \Psi(R)$ a two-sided cone $C(r)$ in $T_r(M)$, given by two linearly independent vectors $w_1(r), w_2(r)$:

$$C(r) = \{v \in T_r(M) | v = a_1 \cdot w_1(r) + a_2 \cdot w_2(r) \text{ with } a_1 \cdot a_2 \geq 0\}.$$

Continuity of C means that w_1 and w_2 depend continuously on r. Let O_r denote the zero vector in $T_r(M)$. An unstable cone field is a continuous cone field on $R \cap \Psi(R)$ such that

– for each $r \in R \cap \Psi(R) \cap \Psi^{-1}(R)$,
$\overline{(d\Psi)(C(r))} \subset \text{Int} \, (C(\Psi(r))) \cup \{O_r\}$,

- for each $r \in R \cap \Psi(R) \cap \Psi^{-1}(R)$ and $v \in C(r)$,
  $\|d\Psi(v)\| \geq \nu \|v\|$, for some $\nu > 1$, where both norms are taken with respect to the basis e_1, e_2.

Below we construct such an unstable cone field. From the existence of such a cone field it follows that there is a continuous direction field $E^u(r)$, defined for $r \in \bigcap_{i \geq 0} \Psi^i(R)$, such that

- $E^u(r) \subset C(r)$,

- $d\Psi$ maps $E^u(r)$ to $E^u(\Psi(r))$,

- for some $\nu > 1$, and all $v \in E^u(r)$ with $r \in \bigcap_{i \geq -1} \Psi^i(R)$, $\|d\Psi(v)\| \geq \nu \cdot \|v\|$.

E^u, restricted to Λ, is obtained by taking the intersections of the forward images of the cone field C under $d\Psi$.

Replacing Ψ by Ψ^{-1}, one constructs in the same way a stable cone field and the direction field E^s, which is invariant under and contracted by $d\Psi$. Then $T_\Lambda(M) = E^u_\Lambda \oplus E^s_\Lambda$ is the required splitting for hyperbolicity—see Appendix 1 and Moser [M,1973a].

Now we come to the construction of the unstable cone field on $R \cap \Psi(R)$. In the component of $R \cap \Psi(R)$ containing p we simply take cones around e_2 extending $45°$ to both sides. In the other component of $R \cap \Psi(R)$ there is (assuming R and $\Psi(R)$ sufficiently thin) an angle α, smaller than $90°$, so that for each point r in that component of $R \cap \Psi(R)$, the cone around $e_2(r)$, extending over an angle α to both sides, contains $e'_2(r)$ in its interior. This is due to the fact that $e_1(r)$ and $e'_2(r)$ are linearly independent. The unstable cone field C is just defined as the field of cones, centred on e_2 and extending $45°$, respectively α, to both sides of e_2 depending on the component of $R \cap \Psi(R)$.

In order to show that this cone field has the required properties, we introduce constants A, B and B' so that: whenever $r \in R \cap \Psi(R)$ and $v = v_1 \cdot e_1(r) + v_2 \cdot e_2(r) \in T_r(M)$ then for $v \in C(r)$ we have $|v_1| \leq A \cdot |v_2|$, on the other hand whenever $|v_1| \leq B \cdot |v_2|$, then $v \in C(r)$; whenever $v = v'_1 \cdot e'_1(r) + v'_2 \cdot e'_2(r)$ and $|v'_1| \leq B' \cdot |v'_2|$, then $v \in C(r)$. If N is so big that $\left|\frac{\lambda}{\sigma}\right|^N \cdot A < \min(B, B')$, then $d\Psi$ maps cones to the interior of cones. Also for N sufficiently big, the lengths of vectors in our cones are strictly increased by $d\Psi$.

So for N big enough our cone field has the required properties. But the cone field was constructed *after* choosing N, Ψ is defined in terms of $N(\Psi = \varphi^N)$, and the domain of the cone field is defined in terms of Ψ: domain $= R \cap \Psi(R)$. However, *the way to raise N is to make R thinner*. This decreases the domain where the cone field has to be defined. The fact that Ψ changes from φ^N to $\varphi^{N'}$, $N' > N$, has no influence on the arguments: the vector fields e_1, e_2, e_1' and e_2' do not change. So R and N may be ajusted afterwards. This completes the proof of the hyperbolicity of Λ. □

Observe that we proved slightly more: there are vector fields $e^u \in e^u_\Lambda$ and $e^s \in e^s_\Lambda$ and a constant $\nu > 1$, such that for all $r \in \Lambda$, $\|d\Psi(e^u(r))\| \geq \nu \cdot \|e^u(r)\|$ and $\|d\Psi(e^s(r))\| \leq \nu^{-1} \cdot \|e^s(r)\|$.

Now we come to the second subject of this section. The cone fields just constructed will now be used to construct the stable and the unstable foliation. We only describe the construction of the unstable foliation. First we present the definition.

An *unstable foliation* for $\Lambda = \bigcap_{i \in \mathbf{Z}} \Psi^i(R)$ is a foliation $\mathcal{F}^u$ of a *neighbourhood* of Λ (here we take $R \cap \Psi(R)$) such that

(a) for each $r \in \Lambda$, $\mathcal{F}^u(r)$, the leaf of $\mathcal{F}^u$ containing r, is tangent to $E^u(r)$,

(b) for each r, sufficiently near Λ, $\Psi(\mathcal{F}^u(r)) \supset \mathcal{F}^u(\Psi(r))$.

We require the tangent directions of leaves of $\mathcal{F}^u$ to vary continuously.

CONSTRUCTION OF THE UNSTABLE FOLIATION: We recall the relative positions of $R, \Psi(R)$, and $\Psi^{-1}(R)$ in Figure 2.10.

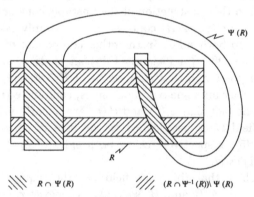

$$\text{\NNN}\quad R \cap \Psi(R) \qquad\qquad \text{////}\quad (R \cap \Psi^{-1}(R)) \setminus \Psi(R)$$

Figure 2.10

We take a C^2 foliation $\tilde{\mathcal{F}}^u$ (not yet the unstable foliation) on $(\Psi(R) \cup \Psi^{-1}(R)) \cap R$ so that:

(a) in $R \cap \Psi(R)$ the tangent directions of leave are contained in the unstable cones,

(b) the images under Ψ of leaves in $(R \cap \Psi^{-1}(R)) \setminus \Psi(R)$ have tangent directions contained in the unstable cones,

(c) the four arcs of $\partial R \cap \Psi^{-1}(R)$ are leaves of $\tilde{\mathcal{F}}^u$, the union of these four arcs denoted by e_0,

(d) the four arcs of $\partial(\Psi(R)) \cap R$ are leaves of $\tilde{\mathcal{F}}^u$, the union of these four arcs is denoted by e_1,

(e) Ψ maps leaves of $\tilde{\mathcal{F}}^u$ near e_0 to leaves of $\tilde{\mathcal{F}}^u$ near e_1.

Since all the cones of the unstable cone field are centred around the vertical vector field e_2 and contain (where defined) e_2', it is clear that such a foliation $\tilde{\mathcal{F}}^u$ exists.

For foliations as described above we define an operator Ψ_* as follows:

in $(R \cap \Psi^{-1}(R)) \setminus \Psi(R)$, the leaves of $\tilde{\mathcal{F}}^u$ and $\Psi_*(\tilde{\mathcal{F}}^u)$ are the same;

in $(R \cap \Psi(R))$, the leaves of $\Psi_*(\tilde{\mathcal{F}}^u)$ are connected components of Ψ-images of leaves of $\tilde{\mathcal{F}}^u$ intersected with $(R \cap \Psi(R))$.

Due to the above conditions (c), (d) and (e), $\Psi_*(\tilde{\mathcal{F}}^u)$ is also C^2. From the invariant manifold theory it follows that the limit

$$\lim_{i \to \infty} \Psi_*^i(\tilde{\mathcal{F}}^u) = \mathcal{F}^u$$

exists. This limit depends on the choice of the "initial foliation" $\tilde{\mathcal{F}}^u$. The limit is C^1; if however φ is C^3 then this limit is $C^{1+\varepsilon}$; see Appendix 1.

Observe that we can extend our vector fields e^u and e^s in E^u, respectively E^s, to tangent vector fields of $\mathcal{F}^u$ and $\mathcal{F}^s$ ($\mathcal{F}^s$ is just an unstable foliation for Ψ^{-1}) so that for some constant $\tilde{\nu} > 1$, and all $r \in R \cap \Psi(R) \cap \Psi^{-1}(R)$,

$$\|d\Psi(e^u(r))\| \geq \tilde{\nu} \cdot \|e^u(r)\|$$

and

$$\|d\Psi(e^s(r))\| \leq \tilde{\nu}^{-1} \cdot \|e^s(r)\|.$$

Stable and unstable foliations can be constructed for any basic set of a C^1 diffeomorphism in dimension 2 [**M**,1973b]. In higher dimensions the existence of such foliations constitutes an interesting open problem except when the basic set is zero-dimensional [**P**,1983]; for futher details see Appendix 1. We stress that in this comment we are insisting on foliations defined in a neighbourhood of the basic set! Sometimes in the literature one also refers to *stable* (or *unstable*) *foliation* as the set of stable manifolds ("leaves") through the points of the basic set, and in this sense the foliation always exist for any hyperbolic set; see Appendix 1.

§4 The maximal invariant subset of R – structure

We divide Λ, the maximal invariant subset of R, into *blocks*. For each sequence $A = (a_{-k}, a_{-k+1}, \cdots, a_{k-1}, a_k)$ with $a_i = 0$ or 1, we define the A-block as $\Lambda_A = \{r \in \Lambda | \Psi^i(r) \in \underline{a}_i$ for $i = -k, \cdots, k\}$; we call k the *radius* of A. The *diameter* of Λ_A is just the supremum of the distances between points in Λ_A, measured in $\mathbb{R}^2$.

As we saw in the last section, expansions and contractions of vectors along unstable, respectively stable, foliations are at least by a factor $\tilde{\nu} > 1$, respectively $\tilde{\nu}^{-1}$. Let c be the maximal length of a component of a stable or unstable leaf in $R \cap \Psi(R)$.

PROPOSITION 1. *Let Λ_A be an A-block and let A have radius k. Then the diameter of Λ_A is at most $2 \cdot c \cdot \tilde{\nu}^{-k}$ (for the definition of c, $\tilde{\nu}$ see above).*

PROOF: For any two points p, p' in the same component of $R \cap \Psi(R)$ there are unique arcs $\ell^u(p, p'')$ and $\ell^s(p', p'')$ in leaves of $\mathcal{F}^u$, respectively $\mathcal{F}^s$, joining p, respectively p', and the intersection p'' of the unstable leaf through p and the stable leaf through p'. (Figure 2.11). When p, p' are both in Λ_A, then this whole configuration will remain in the same component of $R \cap \Psi(R)$ when we apply $\Psi^i, i = -k, \cdots, +k$. This implies that the lengths of $\ell^u(p, p'')$ and $\ell^s(p', p'')$ are both at most $c \cdot \tilde{\nu}^{-k}$. This implies the proposition. $\square$

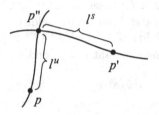

Figure 2.11

From the above proposition and the result of Section 2, we obtain the following

THEOREM 1. *The size of an A-block Λ_A goes to zero as the radius of A goes to infinity. For each infinite sequence $\cdots, a_{-2}, a_{-1}, a_0, a_1, a_2, \cdots, a_i = 0$ or 1, there is exactly one point $r \in \Lambda$ such that $\Psi^i(r) \in \underline{a}_i$ for all i. There is a homeomorphism $h \colon \Lambda \to (\mathbf{Z}_2)^{\mathbf{Z}}$ (product topology on $(\mathbf{Z}_2)^{\mathbf{Z}}$) such that for $r \in \Lambda$, $h(r) = \cdots, a_{-2}, a_{-1}, a_0, a_1, a_2, \cdots$ with $\Psi^i(r) \in \underline{a}_i$. If $\sigma \colon (\mathbf{Z}_2)^{\mathbf{Z}} \to (\mathbf{Z}_2)^{\mathbf{Z}}$ is the shift operator, i.e. $\sigma(\{a_i\}_{i \in \mathbf{Z}} = \{a'_i\}_{i \in \mathbf{Z}}$ with $a'_i = a_{i+1}$, then*

$$
\begin{array}{ccc}
\Lambda & \xrightarrow{\Psi|\Lambda} & \Lambda \\
\downarrow h & & \downarrow h \\
(\mathbf{Z}_2)^{\mathbf{Z}} & \xrightarrow{\sigma} & (\mathbf{Z}_2)^{\mathbf{Z}}
\end{array}
$$

commutes.

REMARK 1: It follows from the above theorem and its proof that if $\tilde{\Psi}$ is C^1−close to Ψ, then the same conclusions hold for the maximal invariant set $\tilde{\Lambda}$ of $\tilde{\Psi}$ in R. Namely, if C is an unstable cone field for Ψ whose domain is slightly extended beyond $R \cap \Psi(R)$, we conclude that C is also an unstable cone field for $\tilde{\Psi}$ if $\tilde{\Psi}$ is sufficiently C^1 close to Ψ. Then all the above arguments apply, with the obvious modifications, to $\tilde{\Psi}$. This implies that for such $\tilde{\Psi}$ and $\tilde{\Lambda}$, there is a conjugacy $H : \Lambda \to \tilde{\Lambda}$, i.e. a homeomorphism such that the diagram below commutes:

$$
\begin{array}{ccc}
\Lambda & \xrightarrow{\Psi|\Lambda} & \Lambda \\
H \downarrow & & \downarrow H \\
\tilde{\Lambda} & \xrightarrow{\tilde{\Psi}|\tilde{\Lambda}} & \tilde{\Lambda}
\end{array}
$$

REMARK 2: The periodic points are dense in Λ. This follows from the corresponding statement for $(\mathbf{Z}_2)^{\mathbf{Z}}$ and σ: any finite sequence can be completed to an infinite periodic sequence. It is clear that all these periodic points are of saddle type (one expanding and one contracting direction). We add one more observation about these periodic orbits to be used in the next chapters. In general we say that a fixed point p of a diffeomorphism φ is *dissipative* if $|\det(d\varphi)(p)| < 1$. The same applies to periodic points, say of period k: just replace φ by φ^k. Now if the fixed point p with which we started this chapter (see Section 1) is dissipative, then, for R sufficiently thin (or N big), all the periodic points in Λ will be dissipative. If $p' \in \Lambda$ is a periodic point of Ψ, and hence of φ, and if R is thin then p' has most points of its orbit (under iteration of φ) in a small neighbourhood of p. The dissipativeness then follows from

$$
\det(d\varphi^k)_{p'} = \prod_{i=0}^{k-1} \det(d\varphi)_{\varphi^i(p')}.
$$

REMARK 3: It also follows from the above constructions that Λ has "local product structure" in the sense that if $r, r' \in \Lambda$ are in the same component of $R \cap \Psi(R)$, then the intersection of the stable leaf $\mathcal{F}^s(r)$ through r and the unstable leaf $\mathcal{F}^u(r')$ through r' also belongs to Λ. In fact, if $h(r) = \{a_i\}_{i \in \mathbf{Z}}$ and $h(r') = \{a'_i\}_{i \in \mathbf{Z}}$, then $\mathcal{F}^s(r) \cap \Lambda$ corresponds to the sequences

$$
\{\{b_i\}_{i \in \mathbf{Z}} | b_i = a_i \text{ for } i \geq 0\}
$$

and $\mathcal{F}^u(r') \cap \Lambda$ corresponds to the sequences

$$
\{\{b'_i\}_{i \in \mathbf{Z}} | b'_i = a'_i \text{ for } i \leq 0\}.
$$

Since r, r' are in the same component of $R \cap \Psi(R)$, $a_0 = a'_0$, so that the point in $\mathcal{F}^s(r) \cap \mathcal{F}^u(r')$ corresponds to the sequence

$$
\cdots, a'_{-2}, a'_{-1}, a'_0 = a_0, a_1, a_2, \cdots.
$$

REMARK 4: A point $r \in \Lambda$ has a *chaotic orbit* if its symbolic sequence $h(r) = \{a_i\}$ is not eventually periodic, i.e., for no n, $a_n, a_{n+1}, \ldots$ is periodic. So in Λ, orbits are either asymptotically periodic or chaotic.

§5 Conclusions for the dynamics near a transverse homoclinic orbit

We return to the diffeomorphism φ (see Section 1) and discuss the consequences of the results we have obtained in Sections 2 to 4 for $\Psi = \varphi^N$. We have analysed the maximal invariant subset Λ in R under the map Ψ; what about the map φ?

This set Λ is contained in $R \cap \Psi(R)$ whose components are denoted by $\underline{0}$ and $\underline{1}$. A corresponding invariant set for φ is defined as $\hat{\Lambda} = \bigcup_{i=0}^{N-1} \varphi^i(\Lambda)$.

PROPOSITION 1. *The set $\hat{\Lambda}$, as defined above, is the disjoint union of $\{p\}$, $\Lambda - \{p\}$, $\varphi(\Lambda - \{p\}), \cdots, \varphi^{N-1}(\Lambda - \{p\})$.*

PROOF: (Figure 2.12). We only have to show that for $0 < i < N$, $\Lambda \cap \varphi^{-i}(\Lambda) = \{p\}$. In fact let $r \in \Lambda$, and $\varphi^i(r) \in \Lambda$ for some $0 < i < N$. Then $r \notin \underline{1}$ and also $\varphi^{k \cdot N}(r) = \Psi^k(r)$ has the same properties, i.e. $\Psi^k(r) \in \Lambda$ and $\varphi^i(\Psi^k(r)) \in \Lambda$. This implies that $\Psi^k(r) \in \underline{0}$ for all k and hence that $r = p$. This proves the proposition. □

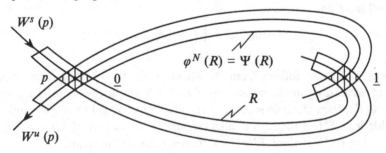

Figure 2.12

It is clear that $\hat{\Lambda}$ is a hyperbolic set for φ, and that the periodic orbits are dense in $\hat{\Lambda}$. As we have observed before, a (transverse) homoclinic orbit implies great complexity of the patterns formed by the corresponding separatrices. In this direction we can prove the following

PROPOSITION 2. *In the above situation, $\hat{\Lambda}$ is contained in the closure of both the stable and the unstable manifold of p.*

PROOF: Since the periodic orbits are dense in $\hat{\Lambda}$ it is enough to prove that each periodic point of $\hat{\Lambda}$ is contained in $\overline{W^s(p)}$ (and in $\overline{W^u(p)}$). Since $\overline{W^s(p)}$ is invariant under φ, it is enough to prove that the periodic points of Λ are in $\overline{W^s(p)}$. For any periodic point $r \in \Lambda$, the unstable separatrix $W^u(r)$ contains the leaf of the unstable foliation through r and hence intersects $W^s(p)$. Then it follows (iterate φ^{-1}) that this periodic point r is contained in the closure of $W^s(p)$. In the same way one proves that it is contained in the closure of $W^u(p)$. $\qquad\square$

§6 Homoclinic points of periodic orbits

Let again $\varphi: M \to M$ be a diffeomorphism but now with a periodic orbit of period k: $\{p_0, p_1, \cdots, p_{k-1}\}, \varphi(p_i) = p_j$ where $j = i+1 \pmod k$. We assume that this periodic orbit is of saddle type. Stable and unstable manifolds are denoted by $W^s(p_i)$ and $W^u(p_i)$. There are two types of homoclinic orbits, namely intersections of $W^s(p_i)$ and $W^u(p_i)$–they are just homoclinic orbits of a hyperbolic saddle fixed point for φ^k–and intersections of $W^s(p_i)$ and $W^u(p_j)$, $i \neq j$. For $t = j - i$, we then have also intersections of $W^s(p_j)$ and $W^u(p_{j+t})$ etc... This means that we get something like a cycle whose "period" is the smallest number ℓ such that $\ell \cdot t$ is a multiple of k. If the intersection of $W^s(p_i)$ and $W^u(p_j)$ is transverse, so are the intersection of $W^s(p_j)$ and $W^u(p_{j+t})$, of $W^s(p_{j+t})$ and $W^u(p_{j+2t})$, etc... By the λ-lemma [P,1969], (see also Appendix 1), this means that $W^s(p_i)$ is accumulating on $W^s(p_j)$ and hence intersecting $W^u(p_{j+t})$ transversally, hence accumulating on $W^s(p_{j+t})$ etc., so that we finally get a transverse intersection of $W^s(p_i)$ with $W^u(p_i)$ anyway.

An example of this last phenomenon occurs in any generic two–parameter family of diffeomorphisms $\varphi_\mu: \mathbb{R}^2 \to \mathbb{R}^2$, $\mu \in \mathbb{R}^2$, such that $\varphi_0(0) = 0$ and such that $(d\varphi_0)_0$ has eigenvalues $e^{\pm 2\pi i/3}$–this is the subharmonic bifurcation with resonance 1:3 [A,1980]. Stable and unstable separatrices are then as indicated in Figure 2.13.

§7 Transverse homoclinic intersections in arbitrary dimensions

As we remarked in the beginning of this chapter, the results and proofs can be extended to diffeomorphisms $\varphi: M \to M$, where M is an n-dimensional manifold. So one obtains the following

THEOREM 1. Let $\varphi: M \to M$ be a C^1 diffeomorphism with a hyperbolic fixed point p. Let q be a point of transverse intersection of $W^u(p)$ and $W^s(p)$.

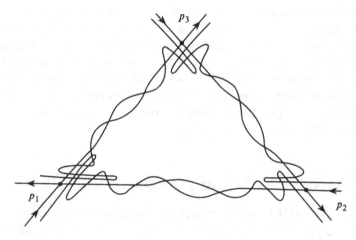

Figure 2.13

Then there is a neighbourhood U of the closure of the orbit $\overline{O(q)} = \bigcup_{i \in \mathbf{Z}} \varphi^i(q)$ such that the maximal invariant set $\hat{\Lambda}$ under φ in U is a nontrivial hyperbolic set (see Appendix 1). Also, there are neighbourhoods V_p and V_q of p and q and there is an integer N, such that the maximal invariant set Λ under φ^N in $V = V_p \cup V_q$ is also a nontrivial hyperbolic set and such that $\varphi^N | \Lambda$ is conjugated with the shift on $(\mathbf{Z}_2)^{\mathbf{Z}}$ as in Section 4.

§8 Historical note

The main ideas in this chapter were developed by Poincaré [**P**,1890], who realized that *transverse homoclinic orbits* are accumulation points of other *homoclinic points*, by G.D. Birkhoff [**B**,1935] who showed that *transverse homoclinic orbits* are accumulation points of *periodic orbits*, and by S. Smale [**S**,1965] who essentially obtained the main theorem of Section 4.

The maximal invariant set in R is often called a horseshoe and the map $\Psi | R$ a horseshoe map. Due to the topology of R^2 there are two types of transverse homoclinic orbits but for both cases the analysis is the same. Our figures refer to the less conventional case in which one does not immediately "see" a horseshoe. In the conventional case one has the dynamic configuration of 2.14(a) instead of the one in 2.14(b).

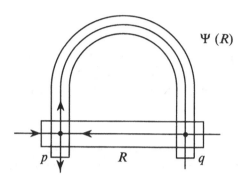

Figure 2.14(a)

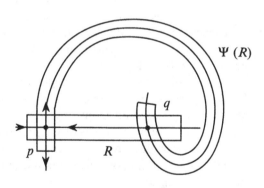

Figure 2.14(b)

CHAPTER 3

HOMOCLINIC TANGENCIES:
CASCADE OF BIFURCATIONS,
SCALING AND QUADRATIC MAPS

In this chapter we begin to present significant results concerning the unfolding of homoclinic tangencies for one-parameter families $\{\varphi_\mu\}$ of diffeomorphisms of a manifold M. As before, for simplicity of presentation, we shall consider here dim $M = 2$. In the last two sections we shall also assume the diffeomorphism to be *locally dissipative*: the product of the eigenvalues at the saddle associated to the homoclinic tangency is smaller than 1. There we show the existence of cascades of period doubling bifurcations of sinks and present a *renormalization* of the family $\{\varphi_\mu\}$ of much importance in our study. The corresponding results in higher dimensions are true if we require the diffeomorphism to have *only one expanding eigenvalue* at the saddle associated to the homoclinic tangency and the *product of any two eigenvalues to have norm smaller than 1*.

The unfolding of a homoclinic tangency yields a great number of changes in the dynamics (bifurcations) as the parameter evolves. In particular the homoclinic tangency is an accumulation point of other homoclinic tangencies. Also many periodic points appear (or disappear) or lose hyperbolicity and change index (i.e., dimension of stable manifold). We have already seen in the last chapter that transverse homoclinic orbits imply the existence of chaotic orbits. When unfolding a homoclinic tangency we expect chaotic dynamics and even chaotic attractors for different maps; see Chapters 6 and 7 for more complete discussions, also in relation with infinitely many coexisting sinks and "apparent chaos". If for some parameter value the map is hyperbolic then there are chaotic orbits but no chaotic dynamics; see Chapters 0 and 5. Finally, we believe that homoclinic tangencies are common among diffeomorphisms whose limit set is nonhyperbolic; see Chapter 7.

Here, in the first three sections, we begin to describe the bifurcation phenomena mentioned above that occur when we unfold a quadratic homoclinic tangency. In the last section we relate this unfolding with the well known family of quadratic maps of the interval $f_\mu(y) = y^2 + \mu$. Remarkably it was for cascades of period doubling bifurcations of this family that Feigenbaum [**F**,1978], [**F**,1979] and Coullet and Tresser [**CT**,1978] (see also [**CE**,1980]) discovered the universality of the limiting speed in the parameter line in which these bifurcations occur.

§1 Cascades of homoclinic tangencies

Let $\phi: M \times R \to M$ be a C^3 map such that $\varphi_\mu(x) = \phi(x, \mu)$ is a diffeomorphism on M for each $\mu \in \mathbb{R}$. We shall denote such a family of diffeomorphisms simply by $\{\varphi_\mu\}$ or just φ_μ. The reason we take the family to be C^3 (and not C^1 or C^2) comes from the discussion of the period doubling bifurcation (or flip) to be presented in the next section of this chapter.

Let us start by studying homoclinic tangencies and their unfoldings. Let $p = p_0$ be a hyperbolic fixed point for φ_0 and let q be a homoclinic tangency associated to p, that is q is a point of tangency between $W^s(p)$ and $W^u(p)$. We assume that $W^s(p)$ and $W^u(p)$ have a quadratic (parabolic) contact at q, and just call q or its orbit $O(q)$ a quadratic homoclinic tangency. (See Figure 3.1).

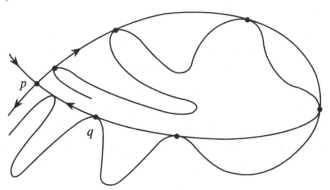

Figure 3.1

We choose local coordinates (x_1, x_2) near q so that we can express the local components of $W^s(p)$ and $W^u(p)$ containing q by

$$W^s(p) = \{(x_1, x_2) \,|\, x_2 = 0\},$$
$$W^u(p) = \{(x_1, x_2) \,|\, x_2 = ax_1^2\}, \tag{1}$$

where $a \neq 0$. Since p is hyperbolic, we have for μ small a unique fixed point p_μ near p and the mapping $\mu \to p_\mu$ is differentiable (implicit function theorem). Also the local components of $W^s(p_\mu)$ and $W^u(p_\mu)$ near q depend differentiably on μ for μ near zero.

Under generic assumptions (see [**NPT**,1983]) there are μ-dependent local coordinates such that $W^s(p_\mu)$ is given by $x_2 = 0$ and $W^u(p_\mu)$ by

$$x_2 = ax_1^2 + b\mu, \qquad a \neq 0 \quad \text{and} \quad b \neq 0. \tag{2}$$

Then we say that the quadratic homoclinic tangency *unfolds generically*.

Taking $a < 0$ and $b > 0$ in (2) above, we get the relative positions of the local components of $W^s(p_\mu)$ and $W^u(p_\mu)$ as shown in Figure 3.2.

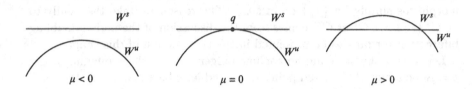

Figure 3.2

THEOREM 1. *Let* $\{\varphi_\mu\}$ *be a one-parameter family of diffeomorphisms with a quadratic homoclinic tangency q at $\mu = 0$ associated to the fixed (periodic) saddle p and suppose it unfolds generically. Then there is a sequence $\mu_n \to 0$ such that φ_{μ_n} has homoclinic tangencies $q_{\mu_n} \to q$ associated to $p_{\mu_n} \to p$.*

PROOF: Let $r = \varphi_0^{-N}(q)$ for some large $N > 0$ and suppose the tangency unfolds into transversal homoclinic points for $\mu > 0$. Given $\mu > 0$ near zero, there are small pieces of "parabolas" (see the above discussion on unfoldings of homoclinic tangencies) $\Gamma_\mu^u \subset W_\mu^u$ near q and $\Gamma_\mu^s \subset W_\mu^s$ near r. We assume that, for $\mu > 0$, their position relative to W_μ^s and W_μ^u near q and r is as indicated in Figure 3.3.

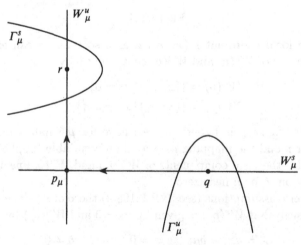

Figure 3.3

Now take $\mu = \hat{\mu}$ arbitrarily small. Clearly, if $n > 0$ is large then $\varphi_{\hat{\mu}}^{-n}(\Gamma_{\hat{\mu}}^s)$ intersects $\Gamma_{\hat{\mu}}^u$. Taking $\mu > 0$ much smaller than $\hat{\mu}$ we have, for the same integer n, that $\varphi_\mu^{-n}(\Gamma_\mu^s) \cap \Gamma_\mu^u = \phi$. Since $\varphi_\mu^{-n}(\Gamma_\mu^s)$ and Γ_μ^u depend C^3 on μ, there is some $0 < \mu_1 < \hat\mu$ for which $\varphi_{\mu_1}^{-n}(\Gamma_{\mu_1}^s)$ and $\Gamma_{\mu_1}^u$ are tangent say at $q_1 \in \Gamma_{\mu_1}^u$. We can repeat the argument for smaller values of $\hat\mu$ and so we can construct the sequences μ_n, q_n as desired, proving the result in the case indicated in Figure 3.3. The reader can easily adapt the argument to other cases, like the one in Figure 3.4. □

REMARK 1: In the proof of the theorem we can take the homoclinic tangencies q_{μ_n} to be of quadratic contact: due to different curvatures, $\varphi_\mu^{-n}(\Gamma_\mu^s)$ and Γ_μ^u have a quadratic contact at their last tangency for decreasing values of μ. One can even show that these homoclinic tangencies unfold generically.

REMARK 2: For the constructions in Chapter 6 it is important to observe that we can even choose the values μ_n in the last theorem so that the branches of $W^s(p_{\mu_n})$ and $W^u(p_{\mu_n})$, i.e. connected components of $W^s(p_{\mu_n}) \setminus \{p_{\mu_n}\}$ and $W^u(p_{\mu_n}) \setminus \{p_{\mu_n}\}$, which have a homoclinic tangency, also have transverse homoclinic intersections.

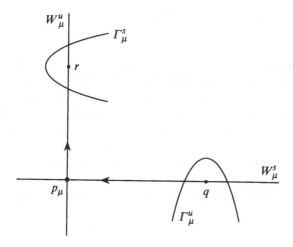

Figure 3.4

§2 Saddle-node and period doubling bifurcations

As we have seen in Chapter 2, transverse homoclinic orbits imply the existence of horseshoes. So, near a homoclinic bifurcation we expect creation or anihilation of horseshoes. In Section 3 we describe the bifurcations due to the formation of such a horseshoe. As a preparation we recall here two of the three generic bifurcations of fixed points in one-parameter families of diffeomorphisms.

Let $\{\varphi_\mu\}_{\mu\in\mathbf{R}}$ be an arc of diffeomorphisms and x_0 a hyperbolic fixed point for φ_0. Then for μ small, φ_μ has a fixed point x_μ, called the continuation of x_0, which is near x_0 for small μ and has the same index as x_0. Thus, for x_{μ_0} to be a bifurcating orbit at least one of the two eigenvalues of $d\varphi_{\mu_0}$ at x_{μ_0} must have norm 1. For generic families, we have three possible cases, as far as the eigenvalues ρ_1 and ρ_2 are concerned:

(a) $\rho_1 = 1$ and $|\rho_2| < 1$ (or $|\rho_2| > 1$),

(b) $\rho_1 = -1$ and $|\rho_2| < 1$ (or $|\rho_2| > 1$),

(c) $\rho_1 = e^{i\theta}$, $\rho_2 = e^{-i\theta}$ for some real $\theta \neq k\pi$, $k \in \mathbf{Z}$.

Case (c), with some further assumptions, corresponds to the so-called *Hopf bifurcation* and it will not be considered here since we will require our mappings to be dissipative (area contracting). In cases (a) and (b), there is a C^3 φ_μ-invariant curve W_μ^c or $W^c(p_\mu)$ which is tangent at $\mu = \mu_0$ to the eigenspace associated with $\rho_1 = 1$ or $\rho_1 = -1$; it is called the centre manifold of φ_μ (see Appendix 1). Thus if we let $f_\mu = \varphi_\mu|W_\mu^c$, we have the following expressions:

$$f_\mu(x) = x + ax^2 + b(\mu - \mu_0) + h \cdot o \cdot t, \tag{1}$$

$$f_\mu(x) = -x + ax^3 + b(\mu - \mu_0)x + h \cdot o \cdot t, \tag{2}$$

$b(0) = 0$. Here (1) corresponds to the first case (a) and (2) to the second case (b); h.o.t stands for higher order terms. We left out the quadratic term in x in (2) because it can be removed by a change of coordinates.

In (1) we take $a \neq 0$ and call the origin a *saddle-node*. We also take $b'(0) \neq 0$ and say that the saddle-node *unfolds generically*. These conditions are clearly satisfied generically. It is easy to see from (1) that f_μ, and thus φ_μ, has two hyperbolic fixed points for $\mu < \mu_0$ and none for $\mu > \mu_0$ or vice versa. If we consider $a > 0$, $b > 0$ and $|\rho_2| < 1$ we have the following unfolding of the saddle-node: a sink and a saddle collapse and then disappear, as shown in Figure 3.5.

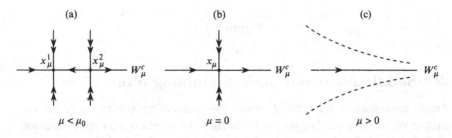

Figure 3.5

The double arrows in the figure mean that the contraction in the normal direction is stronger than along W_μ^c. If we consider, for $\mu \leq \mu_0$, the curves $\mu \to \overline{x}_\mu$, $\mu \to \overline{\overline{x}}_\mu$ of fixed points, we get Figure 3.6.

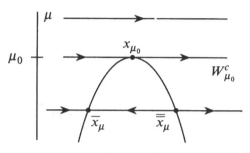

Figure 3.6

Notice that the two curves are differentiable for $|\mu - \mu_0|$ small, $\mu < \mu_0$. If we follow the curve $\mu \to \overline{\overline{x}}_\mu$ for $\mu \nearrow \mu_0$, we can then return along $\mu \to \overline{x}_\mu$ with decreasing values of μ. So the two branches can naturally be oriented as above (or vice versa). In words: if we follow the curve of saddles for increasing values of μ, up to $\mu = \mu_0$, we then return along the curve of sinks for decreasing values of μ. This fact will play a role in the proof of the next theorem.

Now we consider the expression (2) above corresponding to the eigenvalue $\rho_1 = -1$. Similarly to what we have done before in (1), we take $a \neq 0$ (which is a generic condition) and call the orbit a *period doubling bifurcation* (or *flip*); we say that it *unfolds generically* if $b'(0) \neq 0$ (another generic condition!). When $a > 0$ and $b'(0) < 0$, we can easily show that there exists a unique fixed point which is a sink for $\mu < \mu_0$ and a saddle for $\mu > \mu_0$ (both with *negative eigenvalues*); for $\mu > \mu_0$ there is also a *period 2-sink* (with *positive eigenvalues*). Thus the name *period doubling bifurcation*. The results are of course similar in the other cases, where a and $b'(0)$ may have signs different from the ones above. Notice that period doubling which unfolds generically is isolated; the same is true for saddle-nodes. The assumptions and results are also similar for period doubling bifurcations of periodic orbits by just considering the power of the map equal to the period. For instance, a sink of period k may bifurcate into a saddle of period k (both with corresponding negative eigenvalues) and a sink with twice the period (and positive eigenvalues).

For the period doubling bifurcation considered above, if in the set of periodic points we identify points in the same orbit we obtain a topological 1-complex—the curve of sinks for $\mu < \mu_0$, branching off into two topological 1-manifolds: one is formed by the curve of saddles and the other by the curve of sinks with twice the period (see Figure 3.7). Notice that the sink to the left and the saddle to the right both have the same period (and a corresponding negative eigenvalue for df_μ^k, k being the period). This remark will be relevant in the next section.

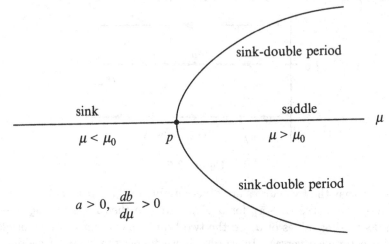

Figure 3.7

§3   Cascades of period doubling bifurcations and sinks

We now discuss the definitions and assumptions of the next theorem showing the existence of many sinks (or sources) and period doubling bifurcations while creating a horseshoe. The sinks that we exhibit in this chapter arise from period doubling bifurcations; they occur for different values of the parameter.

Let R be a rectangle in $\mathbb{R}^2$ and $\{\varphi_\mu\}$ a family of diffeomorphisms of R into $\mathbb{R}^2$ such that

(a) $\varphi_{-1}(R) \cap (R) = \phi$,

(b) $\varphi_\mu | R$ is dissipative (area contracting) for $-1 \leq \mu \leq 1$, that is $|\det(d\varphi_\mu)| < 1$ on R,

(c) φ_1 has periodic points and they are all saddles,

(d) $\varphi_\mu(R) \cap S_1 = \phi$, $\varphi_\mu(R) \cap S_2 = \phi$, $-1 \leq \mu \leq 1$, where S_1, S_2 are two opposite sides in the boundary of R, say the vertical sides,

(e) $\varphi_\mu(T) \cap R = \phi$, $\varphi_\mu(B) \cap R = \phi$, $-1 \leq \mu \leq 1$, where T is the top side of R and B is the bottom side.

In this section we also assume the following *generic* (*residual* or *Baire second category*) condition on the family $\{\varphi_\mu\}$.

(f) φ_μ has at most one nonhyperbolic periodic orbit for each $-1 \leq \mu \leq 1$ and this orbit must correspond either to a saddle-node or to a period doubling bifurcation which unfolds generically. (Because φ_μ is area contracting there is no Hopf bifurcation).

Although we did not formally require φ_1 to be a horseshoe mapping like in Chapter 2, that is precisely the situation we have in mind. In this case, we say that we have an area decreasing family creating a horseshoe, as in Figure 3.8.

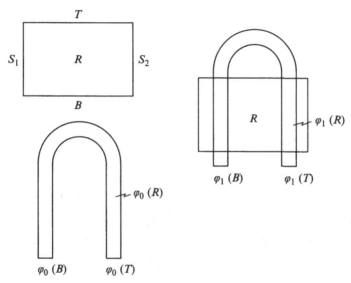

Figure 3.8

Before continuing the discussion, we want to point out that the above conditions are satisfied for a generic unfolding of a homoclinic tangency replacing φ_μ by φ_μ^N, taking $-\delta < \mu < +\delta$ instead of $-1 < \mu < +1$ and choosing R appropriately. Of course, to get the area decreasing property, we assume the determinant of the Jacobian of the map at the fixed (or periodic) saddle with a homoclinic tangency to be less than 1 in absolute value. To see the creation of a horseshoe, let φ_μ be such that

(i) φ_0 has a fixed saddle p and $|\det(d\varphi_0)_p| < 1$,

(ii) there is a generically unfolding homoclinic tangency q associate to p.

We then claim that for each neighbourhood V of q there exists a rectangle $R \subset V$, a number $\delta > 0$ and an integer $N > 0$ such that $\varphi_\mu^N | R$ creates a horseshoe for $-\delta < \mu < \delta$: take R to be a thin rectangle near q and parallel to the local component of W_μ^s as in Figure 3.9.

Let us see why we can choose R, δ and N as wished. First, for μ small, we choose C^1 coordinates linearizing each φ_μ in a fixed neighbourhood of p containing an arc $\ell^s \subset W_0^s$ from p to q; these coordinates may be chosen

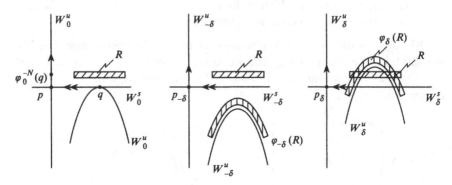

Figure 3.9

to depend continuously on μ (see Appendix 1). We then choose R to be
thin and sufficiently close to W_0^s so that its projection on W_0^u parallel to
W_0^s contains in its interior $\varphi_0^N(q)$ for some large N. Then $\varphi_0^N(R)$ will be a
"curved box" close to an arc in W_0^u near q. For μ near zero, we then have
the situation indicated in the figure. One can then apply arguments similar
to those in Chapter 2 to show that $\varphi_\mu^N|R$ is area decreasing for $-\delta \le \mu \le \delta$
and that $\varphi_\delta^N|R$ has its maximal invariant set hyperbolic with dense subset
of periodic orbits; see also Section 4. In fact, we observe that although the
configuration R, $\varphi_\delta^N(R)$ resembles the situation in Chapter II, the rectangles
considered are quite different: there we had a long rectangle containing p
and q; here the rectangle is contained in a small neighbourhood of q.

We now return to the general discussion about creating horseshoes. For
$\varphi_\mu: \mathbb{R}^2 \to \mathbb{R}^2$ as before, let Per (φ_μ) be the set of periodic orbits of φ_μ
and $P = \{(x, \mu) \mid x \in \text{Per }(\varphi_\mu)\}$. We now define the topological space
$\tilde{P} = P/\sim$, where the equivalence relation $\sim$ is the identification of points in
the same orbit. A component of $\tilde{P}$ through $(O(x), \mu), O(x)$ being the orbit
of the periodic point x, is a continuous curve except at period doubling or
undoubling bifurcations where it branches and looks like Figure 3.10.

THEOREM 1 [**YA**, 1983]. *Let* $\varphi_\mu: \mathbb{R}^2 \to \mathbb{R}^2$ *be a family of orientation
preserving diffeomorphisms satisfying conditions (a) to (f). Then each
$(O(x), 1) \in \tilde{P}$ has a component containing attracting periodic orbits (sinks)
of period $2^n k$ for each $n \ge 0$, where k is the period of x for φ_1.*

REMARK 1: For orientation reversing families $\{\varphi_\mu\}$ one just considers the
squares $\{\varphi_\mu^2\}$; if they satisfy the conditions (a) to (f) we get a corresponding
result.

PROOF: Let $(O(x), 1) \in \tilde{P}$ and assume first that $(d\varphi_1^k)_x$ has *positive eigenvalues*, k being the period of x. By the implicit function theorem, there

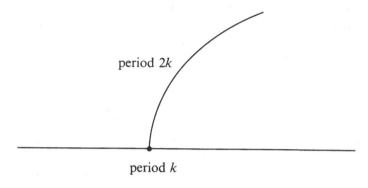

period 2k

period k

Figure 3.10

is a (unique) continuous path Γ in $\tilde{P}$ through $(O(x),1)$ which we follow for decreasing values of μ. We then must reach a bifurcation for otherwise we could follow Γ up to $M \times \{-1\}$ (strictly speaking, Γ is a curve not in $M \times [-1,+1]$ but in $\tilde{P}$, the space of periodic orbits in $M \times [-1,+1]$). In fact, by conditions (d) and (e), the maximal invariant set of φ_μ in R is bounded away from ∂R (periodic points cannot escape through ∂R) and also we cannot terminate Γ in $R \times (-1,1)$ because we can always prolong a path of saddle points. But φ_{-1} has no periodic points in R and so we must reach a bifurcation point and this must correspond to a saddle-node or a period undoubling bifurcation. In both cases we then follow the path of sinks that emanates from the bifurcating orbit for increasing values of μ (see discussions before on saddle-nodes and period doubling bifurcations): see Figure 3.11.

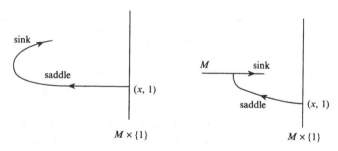

Figure 3.11

In what follows we always prolong Γ at a period doubling or undoubling bifurcation avoiding paths of saddles with corresponding negative eigenvalues (*Moebius paths*) and orient the path positively (for increasing values of μ) if it is a path of sinks and negatively if it is a path of saddles. See Figure 3.12 clarifying the convention. At saddle-nodes the paths are oriented following

the same convention. Now we cannot reach back to $M \times \{1\}$ since φ_1 has
no sinks. Also we cannot have a *cycle*, i.e. Γ cannot return to itself since in
each "turning bifurcating point" there is one path of saddles and one path
of sinks (Moebius paths are not counted here).

We also claim we cannot terminate Γ in $M \times (-1, 1)$ if we go through only
finitely many bifurcating orbits or even infinitely many ones say (x_i, μ_i) with
bounded periods. In fact, in the first case, from our discussions on hyperbolic
and generic bifurcating periodic orbits, we could clearly prolong Γ. In the
second case we can consider a limiting point $(\tilde{x}, \mu)$ of (x_i, μ_i) and argue that
$(\tilde{x}, \mu) \in P$ and then by the genericity assumption (f) on φ_μ, $(\tilde{x}, \mu)$ had to
be locally isolated as a bifurcating orbit of bounded period which is not
the case. Thus we must go through infinitely many bifurcating periodic
orbits with unbounded periods. This can be achieved only if we go through
infinitely many period doubling bifurcations with unbounded periods, which
then clearly implies the result in this case where we started the path at a
saddle $(O(x), 1) \in \tilde{P}$ with positive eigenvalues.

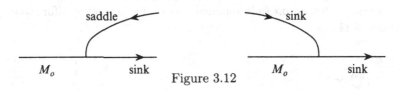

Figure 3.12

Let us now begin with $(O(x), 1) \in \tilde{P}$ such that the eigenvalues of $d\varphi_1^k(x)$
are negative, where k is the period of x, and a path through it in $\tilde{P}$, i.e. a
Moebius path. We will show that the result is also true in this case. Before
we do that, let us again orient in the positive μ-direction paths of sinks and
in the negative μ-direction both paths of saddles with positive eigenvalues
(which we just call paths of saddles) and Moebius paths. So let Γ be a
Moebius path starting at $(O(x), 1)$. As argued before, Γ must go through
bifurcating orbits and the first one must be a period doubling bifurcation.
We then follow the path of saddles of twice the period that emanates from
it. At the next bifurcating orbit we repeat the procedure of prolonging Γ
along the unique non-Moebius path emanating from it. But already at this
point we may get a cycle! That is, a closed oriented path of periodic orbits

not containing any Moebius curves. Figure 3.13 illustrates this possibility.

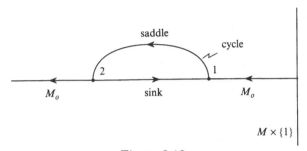

Figure 3.13

If no such cycle appears in Γ we are done. In order to solve the problem with cycles we observe that in each cycle the number of period doublings equals the number of period undoublings. At each period doubling there is an incoming Moebius branch and at each period undoubling there is an outgoing Moebius branch. We then identify in $\tilde{P}$, the space of periodic orbits in $M \times [-1, +1]$, each cycle to one point. The points have now the same (finite) number of ingoing Moebius paths as outgoing Moebius paths. Next we make in this reduced space ($\tilde{P}$ modulo cycles) a path $\tilde{\Gamma}$, starting at $(O(x), 1)$ and consisting entirely of Moebius paths such that

– the orientation of $\tilde{\Gamma}$ agrees with the orientations of the Moebius paths it follows,

– $\tilde{\Gamma}$ passes along each Moebius path at most once.

Such a path can always be continued whenever it reaches a collapsed cycle (the number of ingoing branches equals the number of outgoing branches); such a path cannot end in a periodic orbit in $M \times \{1\}$ since all Moebius paths are oriented towards lower values of μ. So there are two possibilities:

either $\tilde{\Gamma}$ is finite, but then it terminates in a period doubling point not belonging to a cycle; we then proceed as before, and we are done,

or, $\tilde{\Gamma}$ is infinite. In this case the path $\tilde{\Gamma}$ induces a connected graph Γ in $\tilde{P}$, consisting of "arcs of $\tilde{\Gamma}$" and those cycles where $\tilde{\Gamma}$ passes through. Clearly Γ is infinite and hence contains an infinite number of bifurcating periodic orbits. Also here we can use a previous argument to finish the proof. □

§4 Homoclinic tangencies, scaling and quadratic maps

We consider a one-parameter family of diffeomorphisms $\varphi_\mu : M \to M$, M a 2-manifold, which has for $\mu = 0$ a homoclinic tangency. Let p_μ denote the saddle point of φ_μ which is related, for $\mu = 0$, to this tangency. We assume the tangency of $W^u(p_0)$ and $W^s(p_0)$ to be generic (parabolic contact) and also to unfold generically.

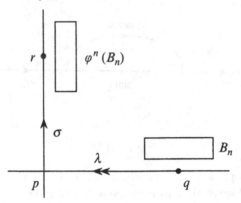

Figure 3.14

Before we go into technicalities, we want to give a heuristic idea of the construction to be described in this section, and its consequences. Near p_0 we take linearizing coordinates x, y so that $\varphi_0(x, y) = (\lambda \cdot x, \sigma \cdot y)$ with $0 < |\lambda| < 1 < |\sigma|$. We assume λ and σ to be positive (otherwise we replace φ_μ by φ_μ^2), and that $\lambda \cdot \sigma < 1$ (if $\lambda \cdot \sigma > 1$ we replace φ_μ by φ_μ^{-1} and if $\lambda \cdot \sigma = 1$ our construction does not work). Let q and r be points on the orbit of tangency in the domain of the linearizing coordinates as indicated in Figure 3.14. So, for some N, $\varphi_0^N(r) = q$. For each sufficiently big n, we take a box B_n near q such that $\varphi_0^n(B_n)$ is a box near r as indicated. We consider $\varphi_\mu^{n+N}(B_n)$, and especially its position relative to B_n. As was already mentioned in Section 3 of this chapter, if one chooses B_n carefully, then, for n sufficiently big, $\varphi_\mu^{n+N}(B_n)$ will cross over B_n so as to create a horseshoe. We shall not only prove this but even show that, after applying n-dependent coordinate transformations to both the (x, y)-variables and the parameter μ (denoting the new variables by $\tilde{x}, \tilde{y}$, and $\tilde{\mu}$), $\varphi_{\tilde{\mu}}^{n+N}$ converges for $n \to \infty$ to the map $\hat{\varphi}_{\tilde{\mu}}$, given by $\hat{\varphi}_{\tilde{\mu}}(\tilde{x}, \tilde{y}) = (\tilde{y}, \tilde{y}^2 + \tilde{\mu})$.

Taking the box B_n in these (n-dependent) coordinates equal to $B = \{(\tilde{x}, \tilde{y}) | |\tilde{x}| \le 3, |\tilde{y}| \le 3\}$ we get the horseshoe formation when $\tilde{\mu}$ decreases from say 4 to -4, at least for n sufficiently big (see Figure 3.15).

Note that this limiting map is not a diffeomorphism any more. This is related to the fact that φ_μ is area contracting at p_μ and hence φ_μ^{n+N}, for $n \to \infty$, becomes more and more area contracting, $\varphi_\mu^{n+N}(B_n)$ tending to be just a curve.

For the limiting map, the value $\tilde{x}$ is unimportant. Restricting to the $\tilde{y}$ variable we have

$$\tilde{y} \mapsto \tilde{y}^2 + \tilde{\mu},$$

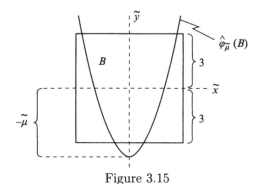

Figure 3.15

which is the well known one-parameter family of quadratic one-dimensional maps, which has been studied e.g. in [**CE**,1980].

This being the limiting map, φ_μ^{n+N} "contains" approximations of this family of quadratic maps and hence exhibits much of its complexity (see Strien [**S**,1981]): hyperbolic sets, period doubling bifurcations and all other phenomena that are persistent under C^r ($r \geq 2$) perturbations.

Because it will be used later, we give here one example of extending a fact about quadratic maps to the one-parameter families like φ_μ. For $\tilde{\mu}$ near zero, the map $\tilde{y} \mapsto \tilde{y}^2 + \tilde{\mu}$ has an attracting fixed point near zero. Let $\mu_n \to 0$ be the sequence of μ-values corresponding to $\tilde{\mu} = 0$ in the different reparametrizations of the μ-variable. Then for n sufficiently big and μ near μ_n, φ_μ^{n+N} has an attracting fixed point.

Now we give a more formal and complete description of the result. First we have to state some extra assumptions on the one-parameter family φ_μ. As we mentioned already we assume that the eigenvalues λ, σ of $(d\varphi_0)_{p_0}$ are positive and satisfy $\lambda \cdot \sigma < 1$. Also we need C^2 linearizing coordinates of φ_μ near p_μ. For this reason we require $\varphi_\mu(x,y)$ to be C^∞ in (μ,x,y). The C^2 linearizing coordinates (μ-*dependent*) then exist, provided that some generic (even open and dense) conditions are satisfied by the eigenvalues λ and σ (see Sternberg [**S**,1958]); they depend continuously, in the C^2 topology, on μ.

THEOREM 1. *For a one-parameter family φ_μ as above, with q a point on the orbit of tangency for $\mu = 0$, there are a constant N and, for each positive integer n, reparametrizations $\mu = M_n(\tilde{\mu})$ of the μ variable and $\tilde{\mu}$-dependent coordinate transformations*

$$(\tilde{x}, \tilde{y}) \mapsto (x, y) = \Psi_{n,\tilde{\mu}}(\tilde{x}, \tilde{y})$$

such that

– for each compact set K in the $(\tilde{\mu}, \tilde{x}, \tilde{y})$–space the images of K under
the maps

$$(\tilde{\mu}, \tilde{x}, \tilde{y}) \mapsto (M_n(\tilde{\mu}), \Psi_{n,\tilde{\mu}}(\tilde{x}, \tilde{y}))$$

converge, for $n \to \infty$, in the μ, x, y space, to $(0, q)$,

– the domains of the maps

$$(\tilde{\mu}, \tilde{x}, \tilde{y}) \mapsto (\tilde{\mu}, (\Psi_{n,\tilde{\mu}}^{-1} \circ \varphi_{M_n(\tilde{\mu})}^{n+N} \circ \Psi_{n,\tilde{\mu}}))$$

converge, for $n \to \infty$, to all of $\mathbb{R}^3$, and the maps converge, in the C^2
topology, to the map

$$(\tilde{\mu}, \tilde{x}, \tilde{y}) \mapsto (\tilde{\mu}, \hat{\varphi}_{\tilde{\mu}}(\tilde{x}, \tilde{y}))$$

with $\hat{\varphi}_{\tilde{\mu}}(\tilde{x}, \tilde{y}) = (\tilde{y}, \tilde{y}^2 + \tilde{\mu})$.

This theorem is an expanded version of a remark in §6.7 (p.336) of
[**GH**,1983]. *Actually, if there are C^k-linearizing coordinates, $k \geq 2$, the
convergence to $\hat{\varphi}_{\tilde{\mu}}$ is in the C^k topology.* This is a consequence of our proof.

PROOF: We start by choosing μ-dependent C^2–linearizing coordinates near
p_μ. We denote them by (x, y). For $\mu = 0$ we have $\varphi_0(x, y) = (\lambda x, \sigma y)$ with
$0 < \lambda < 1 < \sigma$ and $\lambda \cdot \sigma < 1$. Let q be a point in the orbit of tangency in
the "local" stable manifold of p and r such a point in the "local" unstable
manifold of p. See Figure 3.16.

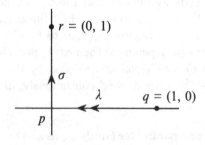

Figure 3.16

By multiplying x, y with constants, we arrange that $q = (1, 0)$ and $r =
(0, 1)$ (see Figure 3.16). Since both r and q are on the orbit of tangency,
there is N such that $\varphi_0^N(r) = q$. For μ near zero we adapt our linearizing
coordinates so that:

– $\varphi_\mu^N(0,1)$ is a local maximum of the y-coordinate restricted to $W^u(p_\mu)$,

– the x-coordinate of $\varphi_\mu^N(0,1)$ is 1.

We reparametrize μ is such a way that the y-coordinate of $\varphi_\mu^N(0,1)$ is μ.

After these preliminary steps we can write φ_μ^N, near $(0,1)$, as

$$(x, 1+y) \mapsto (1,0) + (H_1(\mu, x, y), H_2(\mu, x, y))$$

with

$$H_1(\mu, x, y) = \alpha \cdot y + \tilde{H}_1(\mu, x, y),$$
$$H_2(\mu, x, y) = \beta \cdot y^2 + \mu + \gamma \cdot x + \tilde{H}_2(\mu, x, y)),$$

where α, β, γ are nonzero constants, and where, for $\mu = x = y = 0$,

$$\begin{cases} \tilde{H}_1 = \partial_y \tilde{H}_1 = \partial_\mu \tilde{H}_1 = 0, \\ \tilde{H}_2 = \partial_x \tilde{H}_2 = \partial_y \tilde{H}_2 = \partial_\mu \tilde{H}_2 = \partial_{yy} \tilde{H}_2 = \partial_{y\mu} \tilde{H}_2 = \partial_{\mu\mu} \tilde{H}_2 = 0. \end{cases} \quad (1)$$

The functions H_i and $\tilde{H}_i$ are clearly C^2 since φ_μ is C^∞ and the x, y-coordinates are C^2 in x, y; in fact, instead of $\tilde{H}_2 = \partial_\mu \tilde{H}_2 = \partial_{\mu\mu} \tilde{H}_2 = 0$ and $\partial_y \tilde{H}_2 = \partial_{\mu y} \tilde{H}_2 = 0$, we should put $\tilde{H}_2(\mu, 0, 0) \equiv 0$ and $\partial_y \tilde{H}_2(\mu, 0, 0) \equiv 0$. See Figure 3.17.

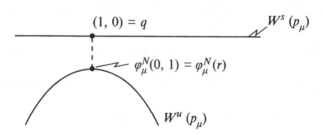

$(1,0) = q$ $W^s(p_\mu)$

$\varphi_\mu^N(0,1) = \varphi_\mu^N(r)$

$W^u(p_\mu)$

Figure 3.17

Next we define an *n-dependent* reparametrization of μ and a μ-dependent coordinate transformation by the following formulas:

$$\mu = \sigma^{-2n} \cdot \bar{\mu} - \gamma \cdot \lambda^n + \sigma^{-n}, \qquad \bar{\mu} = \sigma^{2n} \cdot \mu + \gamma \cdot \lambda^n \cdot \sigma^{2n} - \sigma^n,$$
$$x = 1 + \sigma^{-n} \cdot \bar{x}, \qquad\qquad\quad \bar{x} = \sigma^n \cdot (x - 1),$$
$$y = \sigma^{-n} + \sigma^{-2n} \cdot \bar{y}, \qquad\qquad \bar{y} = \sigma^{2n} \cdot y - \sigma^n.$$

Note that these are *not* yet the final reparametrizations and coordinate transformations but the final ones differ from these just by constant factors, as made explicit at the end of the proof. Also, σ and λ depend on μ and hence on $\bar{\mu}$ although this is not expressed in the above formulas. Finally, note that a fixed box in the $(\bar{x}, \bar{y})$-coordinates, gives for $n \to \infty$ boxes converging to q in the (x, y)-coordinates.

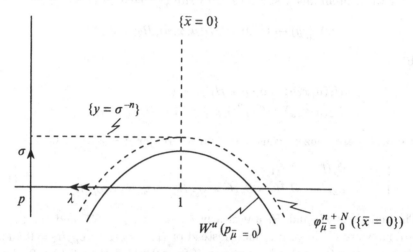

Figure 3.18

We now start with our main calculation: expressing φ_μ^{n+N} in terms of $\bar{\mu}, \bar{x}$ and $\bar{y}$. See Figure 3.18. Let $(\bar{\mu}, \bar{x}, \bar{y})$ denote a point. The (μ, x, y) variables of this point are (see above)

$$\mu = \sigma^{-2n} \cdot \bar{\mu} - \gamma \cdot \lambda^n + \sigma^{-n}, \quad x = 1 + \sigma^{-n} \cdot \bar{x}, \quad y = \sigma^{-n} + \sigma^{-2n} \cdot \bar{y}.$$

After applying φ_μ^n to this point we get as (x, y)-coordinates (μ does not change):

$$x = \lambda^n \cdot (1 + \sigma^{-n}\bar{x}), \qquad y = 1 + \sigma^{-n} \cdot \bar{y}.$$

Next we apply φ_μ^N and find

$$x = 1 + \alpha \cdot \sigma^{-n} \cdot \bar{y} + \tilde{H}_1(\mu, \lambda^n \cdot (1 + \sigma^{-n}\bar{x}), \sigma^{-n} \cdot \bar{y}),$$
$$y = \beta \cdot \sigma^{-2n} \cdot \bar{y}^2 + (\sigma^{-2n}\bar{\mu} - \gamma \cdot \lambda^n + \sigma^{-n})$$
$$+ \gamma \cdot \lambda^n \cdot (1 + \sigma^{-n}\bar{x}) + \tilde{H}_2(\mu, \lambda^n \cdot (1 + \sigma^{-n}), \sigma^{-n} \cdot \bar{y}).$$

Transforming this back to the $(\bar{x}, \bar{y})$-coordinates, and denoting the values of these coordinates of the new point by $\bar{\bar{x}}, \bar{\bar{y}}$ we have

$$\bar{\bar{x}} = \alpha\bar{y} + \sigma^n \cdot \tilde{H}_1(\sigma^{-2n} \cdot \bar{\mu} - \gamma \cdot \lambda^n + \sigma^{-n}, \lambda^n \cdot (1 + \sigma^{-n}\bar{x}), \sigma^{-n} \cdot \bar{y})$$
$$\bar{\bar{y}} = \beta\bar{y}^2 + \bar{\mu} + \gamma \cdot \lambda^n \cdot \sigma^n \cdot \bar{x}$$
$$+ \sigma^{2n} \cdot \tilde{H}_2(\sigma^{-2n} \cdot \bar{\mu} - \gamma \cdot \lambda^n + \sigma^{-n}, \lambda \cdot (1 + \sigma^{-n}\bar{x}), \sigma^{-n} \cdot \bar{y}). \tag{2}$$

Next, we need to show that in the above expression certain parts converge to zero for $n \to \infty$ in the C^2 topology (uniformly on compacta in the $(\bar{\mu}, \bar{x}, \bar{y})$-coordinates).

In the expression for $\bar{\bar{y}}$, the term $\gamma \cdot \lambda^n \cdot \sigma^n \cdot \bar{x}$ goes clearly to zero because $\lambda\sigma < 1$. The terms involving $\tilde{H}_i$ are more complicated. We first observe that when

$$(\bar{\mu}, \bar{x}, \bar{y})$$

remains bounded, the corresponding values of

$$(\mu, x, y - 1)$$

which are substituted in $\tilde{H}_i$ satisfy

$$\mu = O(\sigma^{-n})$$
$$x = O(\lambda^n)$$
$$y - 1 = O(\sigma^{-n}) \tag{3}$$

as n goes to infinity. Next we define

$$\overline{H}_1(\bar{\mu}, \bar{x}, \bar{y}) = \sigma^n \cdot \tilde{H}_1(\sigma^{-2n} \cdot \bar{\mu} - \gamma \cdot \lambda^n + \sigma^{-n}, \lambda^n \cdot (1 + \sigma^{-n} \cdot \bar{x}), \sigma^{-n} \cdot \bar{y}).$$

Then

$$\overline{H}_1(0, 0, 0) = \sigma^n \cdot \tilde{H}_1(-\gamma \cdot \lambda^n + \sigma^{-n}, \lambda^n, 0) = \sigma^n \cdot (O(\lambda^n) + O(\lambda^{2n})),$$

which converges to zero for $n \to \infty$. Next, the first and second order derivatives of $\overline{H}_1(\bar{\mu}, \bar{x}, \bar{y})$ converge to zero (uniformly on compacta); this follows from (1), (3) and $0 < \lambda\sigma < 1$. In fact, the derivatives of $\overline{H}_1$ are easier to estimate than $\overline{H}_1$ itself. This is the way in which one proves that $\overline{H}_1$ goes to zero as announced. The same procedure works for the corresponding expression in the formula for $\bar{\bar{y}}$.

So, for $n \to \infty$, the transformation formulas (2) converge to

$$\begin{pmatrix} \bar{x} \\ \bar{y} \end{pmatrix} \mapsto \begin{pmatrix} \alpha\bar{y} \\ \beta\bar{y}^2 + \bar{\mu} \end{pmatrix}.$$

By the substitution

$$\bar{\mu} = \beta^{-1}\tilde{\mu},$$
$$\bar{x} = \alpha\beta^{-1}\tilde{x},$$
$$\bar{y} = \beta^{-1}\tilde{y},$$

this limiting transformation becomes

$$\begin{pmatrix} \tilde{x} \\ \tilde{y} \end{pmatrix} \mapsto \begin{pmatrix} \tilde{y} \\ \tilde{y}^2 + \tilde{\mu} \end{pmatrix}.$$

Now the theorem is proved: we have the announced transformation as limit of φ_μ^{n+N}, composed with suitable coordinate transformations and reparametrizations of μ. $\square$

We observe that the calculation above, under stronger hypotheses, was carried out independently in [**TY**,1986].

REMARK 1: One can also consider the effect of our (rescaling) coordinate transformations on stable and unstable foliations. Let $\mathcal{F}_\mu^s$ and $\mathcal{F}_\mu^u$ be stable and unstable foliations defined in a neighbourhood of p_μ: i.e. $W^s(p_\mu)$ is a leaf of $\mathcal{F}_\mu^s$, φ_μ maps leaves of $\mathcal{F}_\mu^s$ into such leaves or maps them outside the domain of definition of $\mathcal{F}_\mu^s$, and the tangent directions of $\mathcal{F}_\mu^s$ depend continuously on x, y, and μ; similarly for $\mathcal{F}_\mu^u$. Then we can extend $\mathcal{F}_\mu^s$ by negative iterates of φ_μ and $\mathcal{F}_\mu^u$ by positive iterates of φ_μ till the domains of definition of both these foliations contain the point of tangency of $W^u(p_0)$ and $W^s(p_0)$ in their interior. Next we consider these foliations with respect to the $(\bar{x}, \bar{y}, \bar{\mu})$−or the $(\tilde{x}, \tilde{y}, \tilde{\mu})$-coordinates (depending on n). It turns out, by estimates which are much like the above estimates, that in these coordinates $\mathcal{F}_\mu^s$ and $\mathcal{F}_\mu^u$ converge in the C^2 topology for $n \to \infty$. In the $(\tilde{x}, \tilde{y}, \tilde{\mu})$-coordinates, the limiting leaves of $\mathcal{F}_{\tilde{\mu}}^s$ consist of horizontal curves, the limiting leaves of $\mathcal{F}_{\tilde{\mu}}^u$ consist of the parabolas $\{\tilde{y} = \tilde{x}^2 + a | a \in \mathbb{R}\}$. This topic is discussed in Chapter 6, at the end of Appendix 1 and in Appendix 4.

REMARK 2: If the maps $\varphi_{M_n(\tilde{\mu})}^{n+N}$ are not orientation preserving, then we may replace them by the square $\varphi_{M_n(\tilde{\mu})}^{2(n+N)}$ which is orientable: after reparametrizing according to $\Psi_{n,\tilde{\mu}}$, these maps converge to $(\hat{\varphi}_{\tilde{\mu}})^2$, and, due to renormalization, we know that $(\hat{\varphi}_{\tilde{\mu}})^2$ has, after rescaling $\tilde{\mu}$ and $\bar{x}$, the same properties as $\hat{\varphi}_{\tilde{\mu}}$.

CHAPTER 4

CANTOR SETS IN DYNAMICS AND FRACTAL DIMENSIONS

As already indicated in earlier chapters, the closure of a set of homoclinic intersections is often a Cantor set. In the following chapters, concerning the study of homoclinic bifurcations, we shall have to impose, in the formulations of several results, conditions on such Cantor sets. These conditions will involve numerical invariants like Hausdorff dimension, limit capacity, (local) thickness and denseness, which we discuss in this chapter. When these invariants are nonintegers, one speaks of sets of *fractal dimensions*; often nowadays the name *fractal* is associated to sets whose topological dimension is smaller than their Hausdorff dimension, like the Cantor sets we deal with here.

Since these Cantor sets are not of the most general type, we begin our discussion in the present chapter with the description of "dynamically defined (or regular) Cantor sets"; they form the class of Cantor sets in which we are mainly interested in Dynamical Systems. We prove several results concerning the relations between Hausdorff dimension, limit capacity and (local) thickness and denseness and show that they vary continuously with the maps defining the Cantor sets. These Cantor sets have Hausdorff dimension smaller than 1, and thus their Lebesgue measure is zero. This is not in general the case when the surface diffeomorphism giving rise to the Cantor set is of class C^1, as shown by a counterexample due to Bowen and presented here. Apart from these results that are needed in the next chapters, we derive some additional information concerning the regularity of these Cantor sets, which in our view is of independent interest and bound to play a role in dynamics.

We observe that many of the results in this chapter concern Cantor sets that arise from hyperbolic invariant sets of two–dimensional diffeomorphisms. For hyperbolic sets in higher dimensions, our present knowledge is rather more limited: it is not even known if the definition of local Hausdorff dimension is independent of the initial point.

§1 Dynamically defined Cantor sets

Let $\varphi : M \to M$ be a diffeomorphism of class C^3 of a 2-manifold M with a hyperbolic fixed point p of saddle type which is part of a (nontrivial) basic set Λ. (*Actually, we will see that it is enough to take φ of class C^2; see* *the remark at the end of this section*). By a basic set we mean a compact hyperbolic invariant set with a dense orbit, whose periodic orbits are dense and which is the maximal invariant set in a neighbourhood of it. Nontrivial here means that it does not consist of (finitely many) periodic orbits. As an example, one may think of the "maximal invariant subset of R" related to a transversal homoclinic orbit, as analysed in Chapter 2. For $p \in \Lambda$, the subset $\Lambda \cap W^s(p)$ of $W^s(p)$ is what we want to call a dynamically defined Cantor set. To be more precise, let $\alpha : \mathbb{R} \to W^s(p)$ be a smooth identification, such that $\alpha^{-1} \circ (\varphi|W^s(p)) \circ \alpha$ is a linear contraction (see [S,1957]). Let K be an open and compact neighbourhoord of 0 in $\alpha^{-1}(W^s(p) \cap \Lambda)$: K is called a *dynamically defined Cantor set*. Usually, we even assume that K is obtained by intersecting $\alpha^{-1}(W^s(p) \cap \Lambda)$ with an interval $K_0 \subset R$, containing 0, and whose boundary points are not contained in $\alpha^{-1}(W^s(p) \cap \Lambda)$. We discuss some of the main properties of these Cantor sets.

SCALING: If $\alpha^{-1} \circ (\varphi|W^s(p)) \circ \alpha$ is a linear contraction by λ, which we assume positive, then, since Λ is invariant under φ,

$$\lambda \cdot K = K \cap (\lambda \cdot K_0),$$

where, for $A \subset \mathbb{R}$ and $\lambda \in \mathbb{R}$, $\lambda \cdot A = \{\lambda \cdot a \mid a \in A\}$. This means that the choice of the interval K_0 is not very essential: in each interval $[\lambda \cdot a, a]$ one has the same geometry for the Cantor set K.

EXPANDING STRUCTURE: There is a smooth expanding map $\Psi : K \to K$ with some remarkable properties. We first construct this map. As in Chapter 2 we choose an unstable foliation $\mathcal{F}^u$, defined in a neighbourhood U of Λ. Since the diffeomorphism φ is C^3, this foliation is $C^{1+\varepsilon}$. If the interval K_0 is sufficiently big, then we have a projection π, along leaves of $\mathcal{F}^u$, of a neighbourhood U' of Λ to $\alpha(K_0)$; clearly $\pi(\Lambda) = \alpha(K)$. This projection is in general not unique: one leaf of $\mathcal{F}^u$ may have more than one intersection with $\alpha(K_0)$. Since Λ is totally disconnected, one can still take π, on a small neighbourhood of Λ, continuous and hence differentiable (in fact $C^{1+\varepsilon}$). The derivative of $\pi|(W^s(p) \cap U')$ is bounded and bounded away from zero since the components of $W^s(p) \cap U'$ are leaves of the stable foliation, which is transverse to $\mathcal{F}^u$. For N sufficiently big,

$$\Psi = \alpha^{-1} \circ \pi \circ \varphi^{-N} \circ \alpha : K_0 \to K_0$$

is, where defined, expanding in the sense that the derivative has *norm bigger than 1*. Indeed, the possible contractions in $\pi|(W^s(p) \cap U')$ are compensated by φ^{-N}. From the above construction it follows that $\Psi(K) = K$ (we also denote $\Psi|K$ by Ψ) and that Ψ is $C^{1+\varepsilon}$ in a neighbourhood of K.

Our assumption that K_0 has to be sufficiently big is no real restriction due to the scaling property. The nonuniqueness of Ψ, due to the nonuniqueness of π is still a problem, but this can be bypassed as a consequence of our construction of Markov partitions in Appendix 2.

MARKOV PARTITIONS: For a Cantor set K and an expanding map Ψ as above, we define a Markov partition as a finite set of disjoint intervals $K_1, \ldots, K_k \subset K_0$ such that

- Ψ is defined in a neighbourhood of each K_i, $i \geq 1$,

- K is contained in $\bigcup_{i=1}^{k} K_i$, and the boundary of each K_i is contained in K,

- for each $1 \leq i \leq k$, $\Psi(K_i)$ is an interval, which is the convex hull (in $\mathbb{R}$) of a finite collection of the intervals of the Markov partition,

- for each $1 \leq i \leq k$ and n sufficiently big, $\Psi^n(K \cap K_i) = K$ (this means that $\Psi|K$ is topologically mixing).

For a given Cantor set K as above there are Markov partitions; one can even make the intervals K_i as small as one wishes. This follows from the construction of Markov partitions for basic sets [**B**,1975a]. In the two-dimensional context this construction can be much simplified; see Appendix 2.

Here we only indicate how to make such a Markov partition when our basic set Λ is the horseshoe (see Chapter 1). In this case $W^u(p)$ and $W^s(p)$ are as indicated in Figure 4.1, and $\Lambda = \overline{W^u(p) \cap W^s(p)}$.

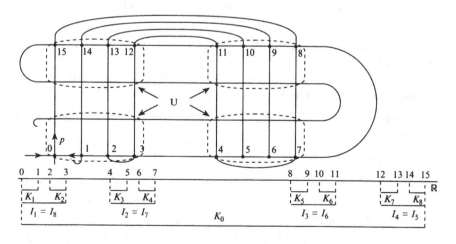

Figure 4.1

In $W^s(p)$ we indicate 16 intersections with $W^u(p)$ (numbered from 0 to 15) they are all in $\alpha(K_0)$. In a "separate copy of $\mathbb{R}$" we indicate the inverse images of these points by α and indicate the intervals $K_1, \ldots, K_8$ of the Markov partition.

In the figure with $W^u(p)$ and $W^s(p)$ we indicate $\alpha(K_0)$ and U, the neighbourhood of Λ in which we assume $\mathcal{F}^u$ to be defined; note that with this choice of U and $\alpha(K_0)$, the projection π (projecting U along fibres of $\mathcal{F}^u$ to $\alpha(K_0)$) is uniquely defined. As expanding map we take

$$\Psi = \alpha^{-1} \circ \pi \circ \varphi^{-1} \circ \alpha.$$

The action of Ψ on the points $0, \ldots, 15$ is then given by

0	1	2	3	4	5	6	7
↓	↓	↓	↓	↓	↓	↓	↓
0	3	4	7	8	11	12	15
↑	↑	↑	↑	↑	↑	↑	↑
15	14	13	12	11	10	9	8

Taking as intervals of the Markov partition $K_1, \ldots, K_8$ as indicated in Figure 4.1, their images are $\Psi(K_i) = I_i$. From this it is simple to verify that $K_1, \ldots, K_8$ form indeed a Markov partition.

Observe that if $\{K_1, \ldots, K_k\}$ a Markov partition of a dynamically defined Cantor set K with expanding map Ψ, one gets a Markov partition with more and shorter intervals by just taking as new intervals connected components of $\Psi^{-1}(K_i) \cap K_j$.

So far we have seen the basic properties of dynamically defined Cantor sets. For the purpose of what follows it is convenient to use these properties as definition.

DEFINITION 1: *A dynamically defined Cantor set* is a Cantor set $K \subset \mathbb{R}$, together with

- a real number λ, the scaling factor, $0 < |\lambda| < 1$, such that λK is a neighbourhood of 0 in K,

- a map $\Psi : K \to K$ having a $C^{1+\varepsilon}$ expansive extension to a neighbourhood of K,

- a Markov partition $\{K_1, \ldots, K_k\}$.

Observe that although in this chapter we will not make use of the scaling property in the definition, this condition will play a key role in Chapter 5.

EXAMPLES: In each of the examples below we *define* the Cantor set by a Markov partition and expanding map. Observe that we can always associate to a Markov partition $\{K_1, \ldots, K_k\}$ and an expanding map Ψ the Cantor set $K = \bigcap_{i=0}^{\infty} \Psi^{-i} (K_1 \cup \cdots \cup K_k)$. Further, we consider only examples where $\Psi|K_i$ is affine, i.e. has constant derivative.

Our first example is the *mid-α-Cantor set*. In this case

$$K_1 = \left[0, \frac{1}{2}(1 - \alpha)\right],$$

$$K_2 = \left[\frac{1}{2}(1 + \alpha), 1\right],$$

and $\Psi|K_i$ maps K_i affinely to $[0, 1]$; the scaling constant can be taken as $\lambda = \frac{1}{2}(1 - \alpha)$. For $\alpha = 1/3$ this is the most well known Cantor set; in any case, for this construction one needs $0 < \alpha < 1$. See Figure 4.2.

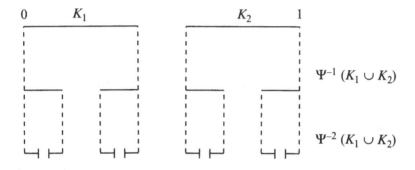

Figure 4.2

The second example, or rather a class of examples, covers the *affine Cantor sets*. They are defined by a sequence of intervals $K_1, \ldots, K_k$ with endpoints K_i^ℓ, K_i^r so that $0 = K_1^\ell < K_1^r < K_2^\ell < K_2^r < K_3^\ell < \cdots < K_k^r$; also $\Psi|K_i$ maps K_i affinely onto $[0, K_k^r]$; as scaling constant one can take $\lambda = K_1^r/K_k^r$. See Figure 4.3.

Finally we define *generalized affine Cantor sets*. They are obtained as the affine Cantor sets, only now the image $\Psi(K_i)$ may be smaller. If we denote the endpoints of K_i as above by K_i^ℓ and K_i^r with $\ldots < K_i^\ell < K_i^r < K_{i+1}^\ell < \ldots$ then $\Psi(K_i)$ should just be an interval of the form $[K_{j_i}^\ell, K_{j_i'}^r]$ with $j_i \le j_i'$. In this case one still has to verify whether Ψ is expanding, whether there is scaling and whether $\Psi^n(K_i \cap K) = K$ for big n. In the special example in Figure 4.4 we have

$$\Psi(K_1 \cap K) = (K_1 \cup K_2) \cap K,$$
$$\Psi(K_2 \cap K) = K,$$
$$\Psi(K_3 \cap K) = (K_2 \cup K_3) \cap K.$$

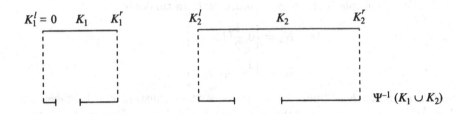

$$\Psi^{-1}\,(K_1 \cup K_2)$$

Figure 4.3

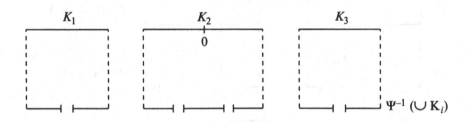

$$\Psi^{-1}\,(\cup\,K_i)$$

Figure 4.4

One sees that Ψ is expanding, $\Psi^n(K_i \cap K) = K$ for $n \geq 2$. In order to have scaling one needs 0 to be a fixed point of the affine map $\Psi|K_2$; here we don't require $0 = K_1^\ell$.

BOUNDED DISTORTION PROPERTY: The above examples are special in the sense that Ψ is affine on each K_i. This is in general not the case. However, as we shall see, the distortions due to the fact that the derivatives of the iterates of Ψ are not locally constant, can be bounded in a very strong sense. It is for these estimates that we require Ψ to be $C^{1+\varepsilon}$.

THEOREM 1. *Let* $K \subset \mathbb{R}$ *be a dynamically defined Cantor set with expanding map* Ψ. *Then, given* $\delta > 0$, *there is* $c(\delta) > 0$ *such that for all* $q, \tilde{q}$ *and* $n \geq 1$ *with*

(a) $|\Psi^n(q) - \Psi^n(\tilde{q})| \leq \delta,$

(b) the interval $[\Psi^i(q), \Psi^i(\tilde{q})]$ contained in the domain of Ψ for all $0 \leq i \leq n - 1$,

we have $|\log|(\Psi^n)'(q)| - \log|(\Psi^n)'(\tilde{q})|| \leq c(\delta)$. Moreover $c(\delta)$ converges to zero when $\delta \to 0$.

PROOF: From the fact that Ψ is expanding it follows that, for some $\sigma > 1$, $|\Psi^i(q) - \Psi^i(\tilde{q})| \leq \delta \cdot \sigma^{i-n}$ for $i \leq n$. Since Ψ is $C^{1+\varepsilon}$ and Ψ' is bounded away from zero, $\log|\Psi'|$ is C^ε. Then

$$|\log|(\Psi^n)'(q)| - \log|(\Psi^n)'(\tilde{q})|| = \left|\sum_{i=0}^{n-1} \log|\Psi'(\Psi^i(q))| - \log|\Psi'(\Psi^i(\tilde{q}))|\right|$$

$$\leq \sum_{i=0}^{n-1} C \left|\Psi^i(q) - \Psi^i(\tilde{q})\right|^\varepsilon \leq \sum_{i=0}^{n-1} C\delta^\varepsilon \cdot \sigma^{\varepsilon(i-n)}$$

$$\leq C\delta^\varepsilon \frac{\sigma^{-\varepsilon}}{1 - \sigma^{-\varepsilon}},$$

for some constant $C > 0$. This proves the theorem by taking $c(\delta) = C\delta^\varepsilon \frac{\sigma^{-\varepsilon}}{1-\sigma^{-\varepsilon}}$. $\square$

Let us finish this section with some comments about the bounded distortion property. First, we present a geometric consequence of it. Let V be some small open interval intersecting K. Since Ψ is topologically mixing (see definition of Markov partition), there is $n \geq 1$ such that $\Psi^n(V \cap K) = K$. Take $q_0, q_1, q_2 \in V \cap K$ close enough to each other, so that the intervals $(\Psi^i(q_0), \Psi^i(q_j))$ are contained in the domain of Ψ for $0 \leq i \leq n - 1$ and $j = 1, 2$. By the mean value theorem, there are $q \in (q_0, q_1)$, $\tilde{q} \in (q_0, q_2)$ such that $|\Psi^n(q_0) - \Psi^n(q_1)| = |q_0 - q_1| \cdot |(\Psi^n)'(q)|$ and $|\Psi^n(q_0) - \Psi^n(q_2)| = |q_0 - q_2| \cdot |(\Psi^n)'(\tilde{q})|$. Then, from the theorem above, we get

$$e^{-c} \frac{|q_0 - q_1|}{|q_0 - q_2|} \leq \frac{|\Psi^n(q_0) - \Psi^n(q_1)|}{|\Psi^n(q_0) - \Psi^n(q_2)|} \leq e^c \frac{|q_0 - q_1|}{|q_0 - q_2|}$$

for some $c > 0$ independent of n, V and the points involved. So Ψ^n essentially preserves ratios of distances between close points: they change but not by more than a uniform, multiplicative constant. This means that, up to a bounded distortion, small parts of K are just reproductions of big parts of K in a smaller scale.

Our second remark concerns the differentiability assumptions in the statement and proof of the theorem above. Clearly we used the assumption that Ψ is $C^{1+\varepsilon}$; in general if Ψ is only C^1 the theorem is not valid. In order to obtain Ψ to be of class $C^{1+\varepsilon}$, it is enough to require our two-dimensional diffeomorphism φ to be of class C^3 since in this case the tangent lines to the leaves of the foliation, and hence the foliation itself, are of class $C^{1+\varepsilon}$. *It is important to observe that Ψ still has the bounded distortion property even when the surface diffeomorphism φ is of class C^2.* The first proof of this fact, which is due to Newhouse and essentially contained in [**N**, 1970; **N**, 1979], follows from the fact that the iterates Ψ^n can be obtained by iterating φ^{-1} n times and *only then* projecting along the leaves of $\mathcal{F}^u$. To illustrate this we consider the basic set Λ to be the horseshoe. Take $\alpha(K_0) \subset W^s(p)$ and U, a neighbourhood of Λ, as in Figure 4.1, and let π denote the projection of U to $\alpha(K_0)$ along the foliation $\mathcal{F}^u$. The fact that this foliation is φ^{-1}-invariant implies that $\pi\varphi^{-1}\pi(x) = \pi\varphi^{-1}(x)$, whenever all these projections are defined. From this, by induction it follows that $(\alpha \circ \Psi \circ \alpha^{-1})^n = \pi \circ \varphi^{-n}$, i.e. $\Psi^n = \alpha^{-1} \circ \pi \circ \varphi^{-n} \circ \alpha$, for all $n \geq 1$ (case $n = 1$ is just the definition of Ψ). *Since, in this expression, the projection π appears only once, its contribution to the distortion of Ψ^n is bounded.* On the other hand the distortion of φ^{-n} can be estimated by the same argument as in the proof of the theorem, since φ^{-1} is, by assumption, C^2 (and so $C^{1+\varepsilon}$). In this way one proves that Ψ has the bounded distortion property. Another proof of this fact consists in showing directly that π, restricted to components of $W^s(p) \cap U$, and so Ψ, is of class $C^{1+\varepsilon}$, even if φ is only C^2—this was communicated to us by Viana.

§2   Numerical invariants of Cantor sets

In this section we define four numerical invariants for Cantor sets, namely Hausdorff dimension, limit capacity, thickness and denseness. Then we discuss the Lebesgue measure of the difference of two Cantor sets in the real line in terms of these invariants and finally we provide some important relations between these invariants when applied to the same Cantor set.

Before we can define Hausdorff dimension, we need to introduce some preliminary notions. Let $K \subset \mathbb{R}$ be a Cantor set and $\mathcal{U} = \{U_i\}_{i \in I}$ a finite covering of K by open intervals in $\mathbb{R}$. We define the diameter diam$(\mathcal{U})$ of $\mathcal{U}$ as the maximum of $\ell(U_i)$, $i \in I$, where $\ell(U_i)$ denotes the length of U_i. Define $H_\alpha(\mathcal{U}) = \sum_{i \in I} \ell_i^\alpha$. Then the *Hausdorff α-measure* of K is

$$m_\alpha(K) = \lim_{\varepsilon \to 0} \left(\inf_{\substack{\mathcal{U} \text{ covers } K \\ \text{diam}(\mathcal{U}) < \varepsilon}} H_\alpha(\mathcal{U}) \right).$$

It is not hard to see that there is a unique number, the *Hausdorff dimension* of K, denoted by $HD(K)$, such that for $\alpha < HD(K)$, $m_\alpha(K) = \infty$ and for $\alpha > HD(K)$, $m_\alpha(K) = 0$.

In order to define the limit capacity, let $N_\varepsilon(K)$, K again a Cantor set in
$\mathbb{R}$, be the minimal number of intervals of length ε needed to cover K. Then
the *limit capacity* of K, denoted by $d(K)$, is defined as

$$d(K) = \limsup_{\varepsilon \to 0} \frac{\log N_\varepsilon(K)}{-\log \varepsilon}.$$

For the mid-α-Cantor set, the first of our examples in the previous section,
one can verify that the Hausdorff dimension and limit capacity are both
equal to $\log 2/(\log 2 - \log(1 - \alpha))$. This will also follow from a more general
result to be discussed later.

Both these notions of Hausdorff dimension and limit capacity can be immediately extended to arbitrary subsets of higher dimensional Euclidean
spaces (or even metric spaces); in the noncompact case one considers countable (instead of finite) coverings.

To define thickness, we consider the *gaps* of K: a gap of K is a connected
component of $\mathbb{R} \setminus K$; a bounded gap is a bounded connected component of
$\mathbb{R} \setminus K$. Let U be any bounded gap and u be a boundary point of U, so
$u \in K$. Let C be the *bridge* of K at u, i.e. the maximal interval in $\mathbb{R}$ such
that

- u is a boundary point of C,

- C contains no point of a gap U' whose length $\ell(U')$ is at least the length
  of U.

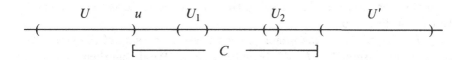

Figure 4.5

In Figure 4.5, U, U', U_1, U_2 are gaps of K, $\ell(U') > \ell(U)$ and C is the
bridge of K at u.

The thickness of K at u is defined as $\tau(K, u) = \ell(C)/\ell(U)$. The *thickness*
of K, denoted by $\tau(K)$, is the infimum over these $\tau(K, u)$ for all boundary
points u of bounded gaps.

Let us provide an equivalent definition of thickness which was actually the one used by Newhouse in [N,1979]. Incidentally and curiously, M. Hall [H,1947] had earlier used the same concept in the context of Number Theory.

Define a *presentation* of the Cantor set K as above to be an ordering $\mathcal{U} = \{U_n\}$ of the bounded gaps of K. For $u \in \partial U_n$, let the $\mathcal{U}$-component of K at u be the connected component C of $I - (U_1 \cup \cdots \cup U_n)$ that contains u. Here, I indicates the minimal (closed) interval of $\mathbb{R}$ containing K. For each such u, denote $\tau(K, \mathcal{U}, u) = \ell(C)/\ell(U_n)$. Then one can check that the thickness of K is given by

$$\tau(K) = \sup_{\mathcal{U}} \inf_{u} \tau(K, \mathcal{U}, u),$$

where the infimum is taken over all boundary points of finite gaps of K and the supremum over all presentations of K. Actually the equality follows from the fact that for any presentation $\mathcal{U} = \{U_n\}$, with $\ell(U_n) \leq \ell(U_m)$ for all $n > m$, the supremum in the formula above is assumed.

We define the *denseness* of K, denoted by $\theta(K)$, as

$$\theta(K) = \inf_{\mathcal{U}} \sup_{u} \tau(K, \mathcal{U}, u).$$

For any two presentations $\mathcal{U} = \{U_n\}$ and $\mathcal{U}' = \{U'_n\}$ of a Cantor set K, we have $\sup_{u} \tau(K, \mathcal{U}, u) \geq \inf_{u} \tau(K, \mathcal{U}', u)$ (take for u the boundary points of U_1), and so we always have $\tau(K) \leq \theta(K)$.

We will show later in the present section that if the thickness is large then the Hausdorff dimension of the Cantor set is also large (i.e., close to 1). We will also provide examples showing that the converse is not generally true. If, however, we replace thickness by denseness then we can state a kind of converse: for a Cantor set in the line, small denseness implies small Hausdorff dimension.

Now we come to the discussion of the Lebesgue measure of the difference of two Cantor sets. Let K_1, K_2 be subsets of $\mathbb{R}$. We define their *difference* as

$$K_1 - K_2 = \{t \in R \mid \exists k_1 \in K_1, k_2 \in K_2, \text{ such that } k_1 - k_2 = t\}.$$

PROPOSITION 1. *Let $K_1, K_2 \subset \mathbb{R}$ be Cantor sets with limit capacity d_1 and d_2. If $d_1 + d_2 < 1$, then the Lebesgue measure of $K_1 - K_2$ is zero.*

PROOF: Let d'_1, d'_2 be numbers such that $d_1 < d'_1$, $d_2 < d'_2$ and $d'_1 + d'_2 < 1$. Then there is an ε_0 such that for $0 < \varepsilon < \varepsilon_0$, K_i can be covered with $\varepsilon^{-d'_i}$ intervals of length ε; this follows directly from the definition of limit capacity. The difference of two intervals of length ε is an interval of length 2ε. So $K_1 - K_2$ is contained in the union of $(\varepsilon^{-d_1'} \cdot \varepsilon^{-d_2'})$ intervals of length 2ε. The total length of these intervals, disregarding overlap, is $2 \cdot \varepsilon^{1-d_1'-d_2'}$. Since $d_1' + d_2' < 1$, this can be made arbitrarily small by choosing ε small. Hence the Lebesgue measure of $K_1 - K_2$ is zero. □

GAP LEMMA. *Let $K_1, K_2 \subset \mathbb{R}$ be Cantor sets with thickness τ_1 and τ_2. If $\tau_1 \cdot \tau_2 > 1$, then one of the following three alternatives occurs: K_1 is contained in a gap of K_2; K_2 is contained in a gap of K_1; $K_1 \cap K_2 \neq \phi$.*

PROOF: We assume that neither of the two Cantor sets is contained in a gap of the other and we assume that $K_1 \cap K_2 = \phi$, and derive a contradiction from this. If U_1, U_2 are bounded gaps of K_1, K_2, we call (U_1, U_2) a *gap pair* if U_2 contains exactly one boundary point of U_1 (and vice versa); U_1 and U_2 are said to be linked in this case. Since neither of the Cantor sets is contained in a gap of the other and since they are disjoint, there is a gap pair. Given such a gap pair (U_1, U_2) we construct:

a point in $K_1 \cap K_2$;

or a different gap pair (U_1', U_2) with $\ell(U_1') < \ell(U_1)$;

or a different gap pair (U_1, U_2') with $\ell(U_2') < \ell(U_2)$.

This leads to a contradiction: even if we don't find a point in $K_1 \cap K_2$ after applying this construction a finite number of times, we get a sequence of gap pairs $(U_1^{(i)}, U_2^{(i)})$ such that $\ell(U_1^{(i)})$ or $\ell(U_2^{(i)})$ decreases and hence, since the sum of all the lengths of bounded gaps is finite, it goes to zero. Assuming $\ell(U_1^{(i)})$ goes to zero, take $q_i \in U_1^{(i)}$: any accumulation point of $\{q_i\}$ belongs to $K_1 \cap K_2$.

Now we come to the announced construction. Let the relative position of U_1 and U_2 be as indicated in Figure 4.6.

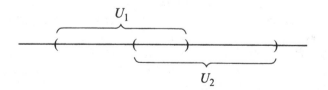

U_1

U_2

Figure 4.6

Let C_j^ℓ and C_j^r be the bridges of K_j at the boundary points of U_j, $j = 1, 2$. Since $\tau_1 \cdot \tau_2 > 1$, $\frac{\ell(C_1^r)}{\ell(U_1)} \cdot \frac{\ell(C_2^\ell)}{\ell(U_2)} > 1$. So $\ell(C_1^r) > \ell(U_2)$ or $\ell(C_2^\ell) > \ell(U_1)$, or both. (See Figure 4.7). Therefore the right endpoint of U_2 is in C_1^r or the left endpoint of U_1 is in C_2^ℓ, or both. Suppose the first. Let u be the right endpoint of U_2. If $u \in K_1$ then we are done, since $u \in K_2$ anyway. If $u \notin K_1$, then u is contained in a gap U_1' of K_1 with $\ell(U_1') < \ell(U_1)$ and (U_1', U_2) is the required gap pair. This completes the proof. $\square$

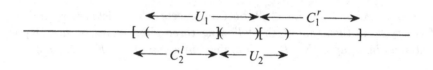

Figure 4.7

REMARK 1: Let now I_1 and I_2 be minimal closed intervals such that $K_1 \subset I_1$ and $K_2 \subset I_2$. We say that K_1 and K_2 are *linked* if I_1 and I_2 are linked. If $\tau(K_1) \cdot \tau(K_2) > 1$ and if K_1 and K_2 are linked, then $K_1 \cap K_2 \neq \phi$ (since neither can K_1 be contained in a gap of K_2 nor K_2 in a gap of K_1). Since being linked is an open condition, it follows that whenever $\tau(K_1) \cdot \tau(K_2) > 1$, then $K_1 - K_2$ has interior points.

THEOREM 1. *Let K_1, K_2 be Cantor sets in $\mathbb{R}$ with Hausdorff dimension h_1, h_2. If $h_1 + h_2 > 1$ then $(K_1 - \lambda K_2)$ has positive Lebesgue measure for almost every $\lambda \in \mathbb{R}$ (in the Lebesgue measure sense).*

Before going into the proof of the theorem, we first observe that from the assumption on h_1, h_2 it follows that $HD(K_1 \times K_2) \geq HD(K_1) + HD(K_2) > 1$ (see Falconer [**F**,1985]). Also, let us see how we can state this result in a similar but slightly different way. For $\lambda \in \mathbb{R}$ take $\theta \in (-\pi/2, +\pi/2)$ such that $\lambda = -\tan \theta$. Let π_θ denote the orthogonal projection of $\mathbb{R}^2$ onto the straight line L_θ which contains $v_\theta = (\cos \theta, \sin \theta)$. If we identify $\mathbb{R}$ with L_θ through $\mathbb{R} \ni x \mapsto x \cdot v_\theta$ then $\pi_\theta(k) = k \cdot v_\theta = \cos \theta \cdot k_1 + \sin \theta \cdot k_2$, for $k = (k_1, k_2) \in \mathbb{R}^2$. By our choice of θ we get $\pi_\theta(K_1 \times K_2) = \cos \theta(K_1 - \lambda K_2)$. Since $\cos \theta \neq 0$ this shows that the theorem above can be rephrased in the following (slightly stronger) form.

THEOREM 2. *Let $K \subset \mathbb{R}^2$ be such that $HD(K) > 1$ and $\pi_\theta : \mathbb{R}^2 \to \mathbb{R}$ be as above. Then $\pi_\theta(K)$ has positive Lebesgue measure for almost every $\theta \in (-\pi/2, +\pi/2)$ (in the Lebesgue measure sense).*

This result was first proved by Marstrand [**M**,1954]. The argument that we present here, which uses ideas from potential theory, is due to Kaufman and can be found in Falconer [**F**,1985].

PROOF: Let $d = HD(K) > 1$. We first assume that $0 < m_d(K) < \infty$ and that for some $C > 0$

$$m_d(K \cap B_r(x)) \leq Cr^d \tag{1}$$

for all $x \in \mathbb{R}^2$ and $0 < r \leq 1$. Let μ be the finite measure on $\mathbb{R}^2$ defined by $\mu(A) = m_d(A \cap K)$, for A a measurable subset of $\mathbb{R}^2$. For $-\pi/2 < \theta < \pi/2$, let us denote by μ_θ the (unique) measure on $\mathbb{R}$ such that $\int f d\mu_\theta = \int (f \circ \pi_\theta) d\mu$ for every continuous function f. The theorem will follow, if we show that the support of μ_θ has positive Lebesgue measure for almost all $\theta \in (-\pi/2, \pi/2)$, since this support is clearly contained in $\pi_\theta(K)$. To do this we use the following fact.

LEMMA 1. *Let η be a finite measure with compact support on $\mathbb{R}$ and $\hat{\eta}(p) = \frac{1}{\sqrt{2\pi}} \int_{-\infty}^{+\infty} e^{-ixp} d\eta(x)$, for $p \in \mathbb{R}$ ($\hat{\eta}$ is the Fourier transform of η). If $0 < \int_{-\infty}^{+\infty} |\hat{\eta}(p)|^2 dp < \infty$ then the support of η has positive Lebesgue measure.*

PROOF OF THE LEMMA: The assumption that $\hat{\eta}$ is square-integrable implies (Plancherel's theorem) that $\varphi(x) = \frac{1}{\sqrt{2\pi}} \int_{-\infty}^{+\infty} e^{ixp} \hat{\eta}(p) dp$ is a well-defined, square-integrable function on $\mathbb{R}$ and $d\eta = \varphi dx$. Moreover $\int_{-\infty}^{+\infty} |\varphi(x)|^2 dx = \int_{-\infty}^{+\infty} |\hat{\eta}(p)|^2 dp > 0$ and so the support of η, which is equal to the support of φ, cannot have Lebesgue measure zero. This proves the lemma. $\square$

Returning to the proof of the theorem we now show that, for almost any $\theta \in (-\pi/2, +\pi/2)$, $\hat{\mu}_\theta$ is square-integrable. From the definitions we have

$$|\hat{\mu}_\theta(p)|^2 = \frac{1}{2\pi} \iint e^{i(y-x)p} d\mu_\theta(x) d\mu_\theta(y)$$
$$= \frac{1}{2\pi} \iint e^{ip(v-u)\cdot v_\theta} d\mu(u) d\mu(v).$$

Then

$$|\hat{\mu}_\theta(p)|^2 + |\hat{\mu}_{\theta+\pi}(p)|^2 = \frac{1}{\pi} \iint \cos(p(v-u) \cdot v_\theta) d\mu(u) d\mu(v)$$

and so

$$\int_0^{2\pi} |\hat{\mu}_\theta(p)|^2 d\theta = \frac{1}{2\pi} \int_0^{2\pi} \iint \cos(p(v-u) \cdot v_\theta) d\mu(u) d\mu(v) d\theta$$
$$= \frac{1}{2\pi} \iint \left(\int_0^{2\pi} \cos(p(v-u) \cdot v_\theta) d\theta \right) d\mu(u) d\mu(v).$$

Note that the integral on θ above does not depend on the direction of $(v-u)$. We now introduce the Bessel function $J_0(z) = \frac{1}{2\pi} \int_0^{2\pi} \cos(z\cos\theta)d\theta$ and write

$$\int_0^{2\pi} |\hat\mu_\theta(p)|^2\, d\theta = \iint J_0(p\,\|v-u\|)d\mu(u)d\mu(v).$$

Integrating on p and using Fubini's theorem we get

$$\int_{-a}^{+a}\int_0^{2\pi} |\hat\mu_\theta(p)|^2\, d\theta dp = \iiint_{-a}^{+a} J_0(p\,\|v-u\|)dp\, d\mu(u)d\mu(v)$$

$$= \iint \left(\int_{-a\|v-u\|}^{+a\|v-u\|} J_0(z)dz \right) \cdot \frac{1}{\|v-u\|}d\mu(u)d\mu(v).$$

Now, it is well known that $\int_{-\infty}^{+\infty} J_0(z)dz$ is convergent. So, in particular, we can write

$$\int_{-a}^{+a}\int_0^{2\pi} |\hat\mu_\theta(p)|^2\, d\theta\, dp \le A \iint \frac{d\mu(u)d\mu(v)}{\|v-u\|}$$

for some $A > 0$ independent of a.

We observe that the integral on the right-hand side is finite. To show this fix $\alpha \in (0,1)$. Then, by condition (1) at the beginning of the proof we have

$$\int \frac{d\mu(v)}{\|u-v\|} = \int_{\|u-v\|\ge 1} \frac{d\mu(v)}{\|u-v\|} + \sum_{n=1}^{\infty} \int_{\alpha^n \le \|u-v\| < \alpha^{n-1}} \frac{d\mu(v)}{\|u-v\|}$$

$$\le \mu(\mathbb{R}^2) + \sum_{n=1}^{\infty} \alpha^{-n}\mu(B_{\alpha^{n-1}}(u))$$

$$\le \mu(\mathbb{R}^2) + \frac{C}{\alpha-\alpha^d}, \quad \text{for all} \quad u \in \mathbb{R}^2,$$

and so

$$\iint \frac{d\mu(u)d\mu(v)}{\|u-v\|} \le \mu(\mathbb{R}^2)\left[\mu(\mathbb{R}^2) + \frac{C}{\alpha-\alpha^d}\right] < \infty.$$

Using Fubini's theorem once more and letting $a \to +\infty$, we get

$$\int_0^{2\pi} \left(\int_{-\infty}^{+\infty} |\hat\mu_\theta(p)|^2\, dp \right) d\theta \le A \iint \frac{d\mu(u)d\mu(v)}{\|v-u\|} < \infty$$

and so $\int_{-\infty}^{+\infty} |\hat\mu_\theta(p)|^2\, dp < \infty$ for almost any $\theta \in (-\pi/2, +\pi/2)$. On the other hand we must have $\int_{-\infty}^{+\infty} |\hat\mu_\theta(p)|^2\, dp > 0$, for all $\theta \in (-\pi/2, +\pi/2)$. In fact if, for some θ, this integral were zero then (recall the notation in the proof of the lemma) $\int_{-\infty}^{+\infty} |\varphi(x)|^2\, dx = 0$, and so $\varphi = 0$ almost everywhere. Since $d\mu_\theta = \varphi dx$ this would imply $\mu_\theta(\mathbb{R}) = \int_{-\infty}^{+\infty} \varphi(x)dx = 0$ and so $\mu(\mathbb{R}^2) = 0$, which contradicts our assumption that the Hausdorff d-measure of K is positive.

This, according to the lemma, proves the theorem, when K has positive, finite d-measure and satisfies condition (1). We now reduce the general case to this one. Take $1 < d' < d$ (so that $m_{d'}(K) = \infty$) and $K' \subset K$ such that $0 < m_{d'}(K') < \infty$ and condition (1) above is satisfied with K' and d' in place of K and d. Such a K' always exists by Theorem 5.6 in Falconer [**F**,1985]. Then the argument above can be carried out with d and K replaced by d' and K', to conclude that $\pi_\theta(K')$ has positive Lebesgue measure for almost all θ, and so the same holds for $\pi_\theta(K)$. $\qquad\square$

REMARK 2: It is worthwhile to point out that if K_1, K_2 are dynamically defined Cantor sets then the Hausdorff d-measure of $K = K_1 \times K_2$ is positive and finite and m_d satisfies condition (1) above, where $d = HD(K_1) + HD(K_2)$. We will further discuss this fact later in this chapter.

Next we come to the relations between the different invariants when applied to the same Cantor set.

PROPOSITION 2. *Let $K \subset \mathbb{R}$ be a Cantor set. Then $d(K) \geq HD(K)$.*

PROOF: For any $d' > d(K)$ and ε sufficiently small, there is a covering of K with $\varepsilon^{-d'}$ intervals of length ε. For such a covering $\mathcal{U}$, and $d'' > d'$, we have $H_{d''}(\mathcal{U}) = \varepsilon^{-d'} \cdot \varepsilon^{d''}$. For ε going to zero, this last expression goes to zero. This means that for any $d'' > d(K)$, the Hausdorff d'' measure of K is zero, and the proposition follows. $\qquad\square$

Observe that this same argument proves a somewhat stronger fact: for any Cantor set $K \subset \mathbb{R}$,

$$HD(K) \leq \liminf_{\varepsilon \to 0} \frac{\log N_\varepsilon(K)}{-\log \varepsilon}.$$

In particular it follows that whenever $HD(K) = d(K)$ then

$$d(K) = \lim_{\varepsilon \to 0} \frac{\log N_\varepsilon(K)}{-\log \varepsilon}$$

(and not just "lim sup"). This is always the case if K is dynamically defined.

THEOREM 3. (See [**T**,1988], [**MM**,1983]). *Let $K \subset R$ be a dynamically defined Cantor set. Then $d(K) = HD(K)$.*

Before giving the formal proof, we want to indicate why the theorem is true in the (easier) case of an affine Cantor set; the proof for the general case is based on the same ideas. So let K be the Cantor set defined by the intervals $K_1, \ldots, K_k$ with endpoints

$$K_1^\ell = 0 < K_1^r < K_2^\ell < \cdots < K_k^r$$

and the affine expanding maps $\Psi_i : K_i \to [0, K_k^r]$. We denote the factor by which distances are multiplied under Ψ_i, by λ_i. So $\lambda_i = \Psi_i' = K_k^r/(K_i^r - K_i^\ell)$. The idea is now to show that both $d(K)$ and $HD(K)$ are equal to the number d for which $\sum \lambda_i^{-d} = 1$. Since $HD(K) \leq d(K)$, we only have to show that $d(K) \leq d$ and $HD(K) \geq d$.

First we indicate why $d(K) \leq d$. Suppose that, for some $\tilde{d} > d$, K can be covered by $\varepsilon^{-\tilde{d}}$ intervals of length ε, whenever $\varepsilon \leq 1$. Then, using the maps Ψ_i^{-1}, we can cover $K \cap K_i$ by $\varepsilon^{-\tilde{d}}$ intervals of length $(\lambda_i^{-1} \cdot \varepsilon)$ for $\varepsilon \leq 1$ or, in other words, by $(\lambda_i^{-\tilde{d}} \cdot \varepsilon^{-\tilde{d}})$ intervals of length ε for $\varepsilon \leq \lambda_i^{-1}$. Since $K = \bigcup(K \cap K_i)$, we can cover K by $((\sum \lambda_i^{-\tilde{d}}) \cdot \varepsilon^{-\tilde{d}})$ intervals of length ε for all $\varepsilon \leq \lambda^{-1}$, where $\lambda = \max \lambda_i$. By induction we find that we can cover K with $((\sum \lambda_i^{-\tilde{d}})^m \cdot \varepsilon^{-\tilde{d}})$ intervals of length ε for all $\varepsilon \leq \lambda^{-m}$.

Since $\tilde{d} > d$, we have $\sum \lambda_i^{-\tilde{d}} < 1$, so for some positive α, $\lambda^{-\alpha} = \sum \lambda_i^{-\tilde{d}}$. Then for $\varepsilon = \lambda^{-m}$ we need no more than

$$\left(\sum \lambda_i^{-\tilde{d}} \right)^m \cdot \varepsilon^{-\tilde{d}} = \varepsilon^{-(\tilde{d}-\alpha)}$$

intervals of length ε to cover K. So $d(K) \leq \tilde{d} - \alpha$ and α is bounded away from zero if $d(K) > d$. Thus $d(K) \leq d$, since $\tilde{d}$ can be taken near $d(K)$.

Next we assume $HD(K) < d$. Then, for $\varepsilon > 0$ and $HD(K) < \tilde{d} < d$ we can find a finite covering $\mathcal{U}$ of K (by intervals) such that $H_{\tilde{d}}(\mathcal{U}) < \varepsilon$. Taking ε sufficiently small, we may assume that each $U_i \in \mathcal{U}$ intersects only one of the intervals $K_1, \ldots, K_k$. The coverings induced by $\mathcal{U}$ on $K \cap K_1, \ldots, K \cap K_k$ are denoted by $\mathcal{U}_i$. Clearly $H_{\tilde{d}}(\mathcal{U}) = \sum H_{\tilde{d}}(\mathcal{U}_i)$. By applying Ψ_i to $\mathcal{U}_i$ we get a covering $\tilde{\mathcal{U}}_i$ of K with $H_{\tilde{d}}(\tilde{\mathcal{U}}_i) = \lambda_i^{\tilde{d}} \cdot H_{\tilde{d}}(\mathcal{U}_i)$. Since $\tilde{d} < d$, we have $\sum \lambda_i^{-\tilde{d}} > 1$ and, hence, at least one of the $H_{\tilde{d}}(\tilde{\mathcal{U}}_i)$ must be smaller than ε. Now $\tilde{\mathcal{U}}_i$ is a new covering of K satisfying again $H_{\tilde{d}}(\tilde{\mathcal{U}}_i) < \varepsilon$, but with *fewer* elements (k, the number of intervals defining K, is at least 2 and each $\mathcal{U}_i$, $i = 1, \ldots, k$, is nonempty). By induction we get such a covering with no elements, which is a contradiction.

PROOF: We begin by describing the structure of the proof. Let $\mathcal{R}^1 = \{K_1, \ldots, K_m\}$ be a Markov partition for K and, for $n \geq 2$, let $\mathcal{R}^n$ denote the set of connected components of $\Psi^{-(n-1)}(K_i)$, $K_i \in \mathcal{R}^1$. For $R \in \mathcal{R}^n$ take $\lambda_{n,R} = \inf |(\Psi^n)'|_R|$ and $\Lambda_{n,R} = \sup |(\Psi^n)'|_R|$. Define $\alpha_n, \beta_n > 0$ by

$$\sum_{R \in \mathcal{R}^n} (\Lambda_{n,R})^{-\alpha_n} = C \text{ and } \sum_{R \in \mathcal{R}^n} (\lambda_{n,R})^{-\beta_n} = 1$$

where C is some properly chosen (big) positive number. We show that, for all $n \geq 1$, we have $HD(K) \geq \alpha_n$ and $d(K) \leq \beta_n$. Finally we prove that $(\beta_n - \alpha_n)_n$ converges to zero as $n \to \infty$; this completes the proof of the theorem.

First we fix the constant C. It follows from the definition of β_n that they are uniformly bounded. Let $\hat{\beta} \geq \beta_n$ for all n. Define $C = \sup |(\Psi^k)'|^{\hat{\beta}}$, where k is such that $\Psi^{k+1}(K_i \cap K) = K$ for all $K_i \in \mathcal{R}^1$. Observe that if $\Psi|K_i$ is onto for all i, then we may take $k = 0$ and so $C = 1$.

Now we prove that $d(K) \leq \beta_n$. Let $\beta > d(K)$. Take $\varepsilon_0 > 0$ so that for $0 < \varepsilon \leq \varepsilon_0$ $N_\varepsilon(K) \leq \varepsilon^{-\beta}$, i.e. there is a covering of K by not more than $\varepsilon^{-\beta}$ intervals of length ε. For every $R \in \mathcal{R}^n$, the inverse images by $(\Psi^n|_R)$ of these intervals form a covering of R by intervals of length at most $\varepsilon\lambda_{n,R}^{-1}$. This means that $N_{\varepsilon\lambda_{n,R}^{-1}}(R) \leq \varepsilon^{-\beta}$ for $0 < \varepsilon \leq \varepsilon_0$, or, in other words, $N_\varepsilon(R) \leq \lambda_{n,R}^{-\beta} \cdot \varepsilon^{-\beta}$ for $0 < \varepsilon \leq \lambda_{n,R}^{-1} \cdot \varepsilon_0$. Then $N_\varepsilon(K) \leq \varepsilon^{-\beta}\left(\sum_{R \in \mathcal{R}^n} \lambda_{n,R}^{-\beta}\right)$ for all $0 < \varepsilon \leq \lambda_n^{-1}\varepsilon_0$, where $\lambda_n = \sup_{R \in \mathcal{R}^n} \lambda_{n,R}$. Repeating the argument we get for all $k \geq 1$

$$N_\varepsilon(K) \leq \varepsilon^{-\beta}\left(\sum_{R \in \mathcal{R}^n} \lambda_{n,R}^{-\beta}\right)^k \quad \text{if} \quad 0 < \varepsilon \leq \lambda_n^{-k}\varepsilon_0.$$

This implies

$$d(K) \leq \beta + \lim_{k \to \infty} \frac{\log\left(\left(\sum_{R \in \mathcal{R}^n} \lambda_{n,R}^{-\beta}\right)^k\right)}{\log(\lambda_n^k \varepsilon_0^{-1})} = \beta + \frac{\log\left(\sum_{R \in \mathcal{R}^n} \lambda_{n,R}^{-\beta}\right)}{\log \lambda_n}$$

and so, making $\beta \to d(K)$,

$$d(K) \leq d(K) + \frac{\log\left(\sum_{R \in \mathcal{R}^n} \lambda_{n,R}^{-d(K)}\right)}{\log \lambda_n}.$$

Since $\lambda_n > 1$ this proves that $\sum_{R \in \mathcal{R}^n} \lambda_{n,R}^{-d(K)} \geq 1$, that is, $d(K) \leq \beta_n$.

Now we derive a contradiction from the assumption that $HD(K) < \alpha_n$. Take $HD(K) < \alpha < \alpha_n$. Then there are finite coverings $\mathcal{U}$ of K with arbitrarily small diameter for which $H_\alpha(\mathcal{U})$ is also arbitrarily small. We assume that every element of $\mathcal{U}$ intersects at most one $R \in \mathcal{R}^n$. This will be the case if we require that $H_\alpha(\mathcal{U}) \leq \varepsilon_0$ for some $\varepsilon_0 = \varepsilon_0(n, \alpha) > 0$. We denote $\mathcal{U}_R = \{U \in \mathcal{U} \,|\, U \cap R \neq \phi\}$. Let, as above, $k \geq 0$ be such that

$\Psi^{k+1}(K_i \cap K) = K$ for all $K_i \in \mathcal{R}^1$. Then, if $H_\alpha(\mathcal{U})$ and hence $\mathrm{diam}(\mathcal{U})$ is sufficiently small, $(\Psi^{n+k} \mid R)(\mathcal{U}_R)$ is a well defined covering of K for all $R \in \mathcal{R}^n$. Note that

$$H_\alpha((\Psi^{n+k}|R)(\mathcal{U}_R)) \leq (\sup |(\Psi^k)'|)^\alpha \cdot \Lambda_{n,R}^\alpha \cdot H_\alpha(\mathcal{U}_R) \leq C \cdot \Lambda_{n,R}^\alpha \cdot H_\alpha(\mathcal{U}_R)$$

(since $\alpha < \alpha_n < \beta_n < \hat{\beta}$). We claim that

$$H_\alpha((\Psi^{n+k}|R_0)((\mathcal{U}_{R_0})) \leq \varepsilon_0$$

for some $R_0 \in \mathcal{R}^n$. Otherwise we would have

$$\begin{aligned}
H_\alpha(\mathcal{U}) = \sum_{R \in \mathcal{R}^n} H_\alpha(\mathcal{U}_R) &\geq C^{-1} \sum_{R \in \mathcal{R}^n} \Lambda_{n,R}^{-\alpha} \cdot H_\alpha((\Psi^{n+k}|R)(\mathcal{U}_R)) \\
&\geq C^{-1}\Big(\sum_{R \in \mathcal{R}^n} \Lambda_{n,R}^{-\alpha} \Big) \cdot \varepsilon_0 \\
&\geq \Big(C^{-1} \sum_{R \in \mathcal{R}^n} \Lambda_{n,R}^{-\alpha} \Big) \cdot H_\alpha(\mathcal{U})
\end{aligned}$$

which is a contradiction, since, by assumption, $\alpha < \alpha_n$ and so

$$\sum_{R \in \mathcal{R}^n} \Lambda_{n,R}^{-\alpha} > C.$$

In this way we construct, from the initial finite covering $\mathcal{U}$, a new covering $\mathcal{U}' = (\Psi^{n+k}|R_0)(\mathcal{U}_{R_0})$, with fewer elements than $\mathcal{U}$ and such that $H_\alpha(\mathcal{U}') \leq \varepsilon_0$. Repeating this argument we eventually obtain a covering of K with no elements at all. This is the required contradiction.

Finally, to prove that $(\beta_n - \alpha_n)_n \to 0$ we first note that, by the bounded distortion property there is $a > 0$, such that $\Lambda_{n,R} \leq a \cdot \lambda_{n,R}$, for all $n \geq 1$ and $R \in \mathcal{R}^n$. Take $\delta_n = \dfrac{\alpha_n \log a + \log C}{-\log a + n \log \lambda}$, where $\lambda = \inf |\Psi'| > 1$. Then

$$\begin{aligned}
\sum_{R \in \mathcal{R}^n} \lambda_{n,R}^{-(\alpha_n + \delta_n)} &\leq a^{(\alpha_n + \delta_n)} \sum_{R \in \mathcal{R}^n} \Lambda_{n,R}^{-\alpha_n} \cdot \Lambda_{n,R}^{-\delta_n} \\
&\leq a^{(\alpha_n + \delta_n)} \cdot \lambda^{-n \cdot \delta_n} \cdot \sum_{R \in \mathcal{R}^n} \Lambda_{n,R}^{-\alpha_n} \\
&= a^{(\alpha_n + \delta_n)} \cdot \lambda^{-n\delta_n} \cdot C = 1,
\end{aligned}$$

by definition of δ_n. It follows that $\beta_n \leq \alpha_n + \delta_n$, i.e.

$$\beta_n - \alpha_n \leq \frac{\alpha_n \log a + \log C}{n \log \lambda - \log a} \leq \frac{HD(K) \cdot \log a + \log C}{n \log \lambda - \log a}.$$

This implies the convergence we have claimed and completes the proof of the theorem. □

The above theorem is a consequence of the regularity of dynamically defined Cantor sets. It makes the propositions on the measure of the difference of two Cantor sets, in terms of limit capacity and Hausdorff dimension, cover, for dynamically defined Cantor sets, almost all cases the exceptions being $d(K_1) + d(K_2) = 1$ and $K_1 - \lambda K_2$ for exceptional values of λ. Before proceeding with our discussion on the relations between the invariants (dimensions) of a Cantor set, let us explore some consequences of the ideas involved in the proof of this theorem.

First we recall that in the heuristic proof we have the following formula for the Hausdorff dimension and the limit capacity. If K is an affine Cantor set (see the examples in the previous section) with Markov partition $\{K_1, \dots, K_k\}$ and λ_i denotes the (constant) value of $|\Psi'|_{K_i}|$, then $HD(K) = d(K) = d$, d being the unique number such that $\sum \lambda_i^{-d} = 1$. We use this formula to compute the precise value of $HD(K) = d(K)$ in a particular case. Take K to be an affine Cantor set with Markov partition $\{K_1, \dots, K_k\}$ such that all the K_i have equal length, say $\beta \cdot \operatorname{diam} K$ for $0 < \beta < 1/k$. Since we are assuming that K is affine (and not just generalized affine), Ψ maps each $K_i \cap K$ onto K, so we must have $\lambda_i = \beta^{-1}$ for all i. Therefore $HD(K) = d(K) = \log k / \log(\beta^{-1})$. For $k = 2$, since $\beta = (1 - \alpha)/2$, we get the formula stated at the beginning of this section. Incidentally, this shows that the dimension of a dynamically defined Cantor set can take any value between 0 and 1. Also, for any $\rho \in (0, 1)$, there are diffeomorphisms exhibiting a saddle point p and a basic set Λ with $p \in \Lambda$ such that $HD(\Lambda \cap W^s(p)) = \rho$.

Our second remark concerns the role played by the bounded distortion property. Although we made use of it in the last part of the proof this is not strictly necessary for the theorem above. In fact this result is still true for Cantor sets defined by expanding maps which are only C^1 (and so may not have this property); see [**T**,1988]. Even more so, if φ is just a C^1 diffeomorphism, still the Cantor sets $W^s(p) \cap \Lambda$ induced by it have their Hausdorff dimension equal to the limit capacity; see [**PV**,1988]. However, by using the bounded distortion property one can give better estimates for the velocity of convergence of α_n and β_n, than would hold if Ψ were just C^1.

Recall that in the proof of the theorem we showed that, for some $b > 0$,

$$\beta_n - \frac{b}{n} \le \alpha_n \le d(K) = HD(K) \le \beta_n \le \alpha_n + \frac{b}{n}, \text{ for all } n \ge 1. \quad (2)$$

We want to explore some important consequences of this estimate. First observe that, denoting $A = \sup |\Psi'|$ and $d = HD(K) = d(K)$,

$$\sum_{R \in \mathcal{R}^n} \lambda_{n,R}^{-d} = \sum_{R \in \mathcal{R}^n} \lambda_{n,R}^{-\beta_n} \cdot \lambda_{n,R}^{\beta_n - d} \leq \sum_{R \in \mathcal{R}^n} \lambda_{n,R}^{-\beta_n} \cdot A^{n(\beta_n - d)}$$

and so

$$\sum_{R \in \mathcal{R}^n} \lambda_{n,R}^{-d} \leq A^b < \infty, \text{ for all } n \geq 1. \tag{3}$$

In a similar way,

$$\sum_{R \in \mathcal{R}^n} \Lambda_{n,R}^{-d} \geq C A^{-b} > 0, \text{ for all } n \geq 1. \tag{4}$$

Using these facts we prove the following proposition.

PROPOSITION 3. Let $K \subset \mathbb{R}$ be a dynamically defined Cantor set and let $d = HD(K)$. Then, $0 < m_d(K) < \infty$. Moreover, there is $c > 0$ such that, for all $x \in K$ and $0 < r \leq 1$,

$$c^{-1} \leq \frac{m_d(B_r(x) \cap K)}{r^d} \leq c. \tag{5}$$

We point out that the bounded distortion property is fundamental here: contrary to the theorem above, this last proposition wouldn't hold in general if Ψ were only C^1.

PROOF: We keep the notations from the proof of the above theorem. Observe that by the mean value theorem and condition (3),

$$H_d(\mathcal{R}^n) = \sum_{R \in \mathcal{R}^n} [\ell(R)]^d \leq \sum_{R \in \mathcal{R}^n} \left(\lambda_{n,R}^{-1} \cdot \ell(K) \right)^d \leq A^b (\ell(K))^d,$$

where $\ell(K)$ denotes the diameter of K. Since $\text{diam}(\mathcal{R}^n) \to 0$ as $n \to \infty$, this proves that

$$m_d(K) \leq A^b (\ell(K))^d < \infty.$$

Proving that $m_d(K)$ is positive requires a little more effort. First, we claim that for some $a_1 > 0$ we have

$$(\ell(U))^d \geq a_1 \cdot \sum_{\substack{R \in \mathcal{R}^n \\ R \cap U \neq \phi}} \Lambda_{n,R}^{-d} \tag{6}$$

for every interval U intersecting K and $n \geq 1$ sufficiently large depending on U. To show this we fix $\alpha > 0$ such that the α-neighbourhood of K is contained in the domain of Ψ. Take $k = k(U) \geq 0$ minimal such that

$$\ell(\Psi^k(U)) \geq \alpha.$$

Let $n > k$. Then $S \in \mathcal{R}^{n-k}$ intersects $\Psi^k(U)$ if and only if $S = \Psi^k(R)$ for some $\mathcal{R} \in \mathcal{R}^n$ intersecting U. Moreover, in such case we have

$$\Lambda_{n,R} = \sup |(\Psi^n)'|_R| \geq \inf |(\Psi^k)'|_R| \cdot \sup |(\Psi^{n-k})'|_S|$$
$$\geq \inf |(\Psi^k)'|_{U \cup R}| \cdot \Lambda_{n-k,S}.$$

On the other hand, by the mean value theorem we have

$$\ell(\Psi^k(U)) \leq \sup |(\Psi^k)'|_U| \cdot \ell(U) \leq \sup |(\Psi^k)'|_{U \cup R}| \cdot \ell(U).$$

Observe that, by construction, $\Psi^j(U \cup R)$ is contained in the domain of Ψ for all $0 \leq j \leq k - 1$. The bounded distortion property implies that

$$\sup |(\Psi^k)'|_{U \cup R}| \leq a \cdot \inf |(\Psi^k)'|_{U \cup R}|,$$

where a is some positive number independent of U, R and k. From all this and the fact that $d \geq \alpha_{n-k}$, we obtain

$$(\ell(U))^d \geq C^{-1} \sum_{\substack{S \in \mathcal{R}^{n-k} \\ S \cap \Psi^k(U) \neq \phi}} \Lambda_{n-k,S}^{-d} \cdot \ell(U)^d$$

$$\geq C^{-1} \sum_{\substack{R \in \mathcal{R}^n \\ R \cap U \neq \phi}} \Lambda_{n-k,\Psi^k(R)}^{-d} \cdot (\ell(\Psi^k(U)) / \sup |(\Psi^k)'|_{U \cup R}|)^d$$

$$\geq C^{-1} \alpha^d \sum_{\substack{R \in \mathcal{R}^n \\ R \cap U \neq \phi}} \Lambda_{n-k,\Psi^k(R)}^{-d} \cdot a^{-d} \cdot (\inf |(\Psi^k)'|_{U \cup R}|)^{-d}$$

$$\geq C^{-1} \alpha^d a^{-d} \sum_{\substack{R \in \mathcal{R}^n \\ R \cap U \neq \phi}} \Lambda_{n,R}^{-d}.$$

This proves the claim with $a_1 = C^{-1} \alpha^d a^{-d}$.

Let now $\mathcal{U}$ be any finite covering of K. Take $n \geq 1$ such that (6) holds for all $U \in \mathcal{U}$. Then, by (4),

$$H_d(\mathcal{U}) = \sum_{U \in \mathcal{U}} (\ell(U))^d \geq \sum_{U \in \mathcal{U}} a_1 \Big(\sum_{\substack{R \in \mathcal{R}^n \\ R \cap U \neq \phi}} \Lambda_{n,R}^{-d} \Big)$$

$$\geq a_1 \sum_{R \in \mathcal{R}^n} \Lambda_{n,R}^{-d} \geq a_1 C A^{-b}.$$

Since $\mathcal{U}$ is arbitrary, this proves

$$m_d(K) \geq a_1 C A^{-b} > 0.$$

Now we deal with the second part of the proposition. To make the argument more transparent we first derive an estimate for the d-measure of the intervals $R \in \mathcal{R}^n$. For some $a_2 > 1$, depending only on K and Ψ, we have

$$a_2^{-1} \leq \frac{m_d(R \cap K)}{(\ell(R))^d} \leq a_2 \tag{7}$$

for all $R \in \mathcal{R}^n$ and $n \geq 1$. To show this we observe that Ψ^{n-1} maps R diffeomorphically onto some $K_i \in \mathcal{R}^1$. From the definition of Hausdorff measure, we have

$$\lambda_{n-1,R}^d \cdot m_d(R \cap K) \leq m_d(K_i \cap K) \leq \Lambda_{n-1,R}^d \cdot m_d(R \cap K).$$

On the other hand, by the mean value theorem, we have

$$\lambda_{n-1,R} \cdot \ell(R) \leq \ell(K_i) \leq \Lambda_{n-1,R} \cdot \ell(R).$$

Finally, by the bounded distortion property, it follows that

$$\Lambda_{n-1,R} \leq a \cdot \lambda_{n-1,R}$$

with $a > 0$ as above depending only on K and Ψ. From all this we get

$$a^{-d} \cdot \frac{m_d(K_i \cap K)}{(\ell(K_i))^d} \leq \frac{m_d(R \cap K)}{(\ell(R))^d} \leq \frac{m_d(K_i \cap K)}{(\ell(K_i))^d} \cdot a^d.$$

Clearly, $\ell(K_i)$ can be uniformly bounded from zero and infinity, so to prove (7) we only need to show that the same holds for $m_d(K_i \cap K)$. The upper bound is trivial since $m_d(K_i \cap K) \leq m_d(K) < \infty$.

The lower bound follows easily from the fact that, for some $k \geq 0$, $\Psi^{k+1}(K_i \cap K) = K$ and so, again by the definition of Hausdorff measure,

$$m_d(K_i \cap K) \geq \left(\sup \left| (\Psi^{k+1})' \right| \right)^{-d} m_d(K) > 0.$$

Now we prove (5). For $x \in K$ and $0 < r \leq 1$, we let $q = q(x,r) \geq 0$ be minimal such that

$$\Psi^q(B_r(x)) \not\subset B_\alpha(\Psi^q(x))$$

where, as before, $\alpha > 0$ is such that the domain of Ψ contains the α-neighbourhood of K. Then, arguing as above with $B_r(x)$ and Ψ^q in the place of R and Ψ^{n-1}, respectively, we obtain

$$a^{-d} \cdot \frac{m_d(\Psi^q(B_r(x)) \cap K)}{\ell(\Psi^q(B_r(x)))} \leq \frac{m_d(B_r(x) \cap K)}{(2r)^d} \leq \frac{m_d(\Psi^q(B_r(x)) \cap K)}{\ell(\Psi^q(B_r(x)))} \cdot a^d.$$

Again, $\ell(\Psi^q(B_r(x)))$ can be easily bounded: by construction

$$\alpha \leq \ell(\Psi^q(B_r(x))) \leq A \cdot \alpha.$$

Since we also have

$$m_d(\Psi^q(B_r(x)) \cap K) \leq m_d(K) < \infty,$$

it is enough to provide a uniform lower bound for the d-measure of $\Psi^q(B_r(x)) \cap K$. To do this, we observe that by the mean value theorem we have

$$B_{r\Lambda_q}(\Psi^q(x)) \supset \Psi^q(B_r(x)) \supset B_{r\lambda_q}(\Psi^q(x)),$$

where $\lambda_q = \inf \left|(\Psi^q)'|_{B_r(x)}\right|$ and $\Lambda_q = \sup \left|(\Psi^q)'|_{B_r(x)}\right|$. Then, by the definition of q,

$$r\Lambda_q \geq \alpha$$

and so, using the bounded distortion property once again, $\Psi^q(B_r(x)) \supset B_{\alpha a^{-1}}(\Psi^q(x))$. Fix $p \geq 1$ such that $\ell(R) < \alpha a^{-1}$ for all $R \in \mathcal{R}^p$. Then, $\Psi^q(B_r(x))$ must contain the interval $R_0 \in \mathcal{R}^p$ that contains $\Psi^q(x)$. Finally, one proves, as we did before for $K_i \in \mathcal{R}^1$, that $m_d(R \cap K) > 0$ for all $R \in \mathcal{R}^p$. It follows that

$$m_d(\Psi^q(B_r(x)) \cap K) \geq m_d(R_0 \cap K) \geq \inf\{m_d(R \cap K) \mid R \in \mathcal{R}^p\} > 0$$

and this completes the proof of the proposition. $\qquad\square$

Finally, we prove a two-dimensional version of this proposition, which had been remarked following the proof of Marstrand's theorem relating Hausdorff dimension and measure of the difference set. It applies to hyperbolic basic sets of diffeomorphisms on surfaces.

PROPOSITION 4. *Let K_1, K_2 be dynamically defined Cantor sets and let $d_1 = HD(K_1)$, $d_2 = HD(K_2)$, $d = d_1 + d_2$ and $K = K_1 \times K_2$ in $\mathbb{R}^2$. Then, for some $c > 0$,*

(a) $0 < m_d(K) < \infty$,

(b) $c^{-1} \leq \frac{m_d(K \cap B_r(x))}{r^d} \leq c$ *for all $x \in K$ and $0 < r \leq 1$.*

PROOF: Take μ to be the product measure $\mu = m_{d_1} \times m_{d_2}$ on K. Clearly, (a) and (b) hold if we replace there m_d by μ. Therefore it is now sufficient to show that μ is equivalent to m_d in the sense that for all Borel subsets $A \subset K$, $\mu(A)/m_d(A)$ is bounded away from zero and infinity. We consider Markov partitions $\mathcal{R}_1, \mathcal{R}_2$ for K_1, K_2 respectively and denote by $\mathcal{R}_i^n$, $i = 1, 2$, the family of connected components of $\Psi_i^{-(n-1)}(L_j)$, $L_j \in \mathcal{R}_i$. We may restrict ourselves to Borel sets of the form

$$A = R_1 \times R_2, \quad R_1 \in \mathcal{R}_1^n, \quad R_2 \in \mathcal{R}_2^n$$

since these sets generate the Borel σ-algebra of K. Let $\mathcal{U} = \{U_{1,j} \times U_{2,j} \mid 1 \leq j \leq m\}$ be any finite covering of $A = R_1 \times R_2$ by cubes. Fix $x_{i,j} \in U_{i,j} \cap R_i$, $i = 1, 2, 1 \leq j \leq m$ (obviously, we may assume $U_{i,j} \cap R_i \neq \emptyset$). Then

$$\begin{aligned}
\mu(U_{1,j} \times U_{2,j}) &= m_{d_1}(U_{1,j}) \times m_{d_2}(U_{2,j}) \\
&\leq m_{d_1}(B_{1,j}) \times m_{d_2}(B_{2,j}) \\
&\leq c_1 c_2 \cdot (\ell(U_{1,j}))^{d_1} \cdot (\ell(U_{2,j}))^{d_2}
\end{aligned}$$

where $B_{i,j}$ denotes the ball in K_i centred in $x_{i,j}$ and with radius $\ell(U_{i,j})$. Therefore

$$\begin{aligned}
\sum_{j=1}^m (\mathrm{diam}(U_{1,j} \times U_{2,j}))^d &\geq \sum_{j=1}^m (\ell(U_{1,j}))^{d_1} \times (\ell(U_{2,j}))^{d_2} \\
&\geq (c_1 c_2)^{-1} \sum_j \mu(U_{1,j} \times U_{2,j}) \\
&\geq (c_1 c_2)^{-1} \mu(A).
\end{aligned}$$

Since $\mathcal{U}$ is arbitrary this proves

$$m_d(A) \geq (c_1 c_2)^{-1} \mu(A).$$

To obtain an inequality in the opposite direction we construct coverings $\mathcal{U}_m$, of $A = R_1 \times R_2$, $m >> n$, as follows. Fix $U_1 \in \mathcal{R}_1^m$, U_1 contained in R_1. For each $x_2 \in R_2$, take $\overline{m}(U_1, x_2)$ maximal such that if $U_2(U_1, x_2)$ denotes the element of $\mathcal{R}_2^{\overline{m}(U_1, x_2)}$ containing x_2, then $\ell(U_2(U_1, x_2)) \geq \ell(U_1)$. Clearly, $\{U_2(U_1, x_2) : x_2 \in R_2\}$ contains a finite covering of R_2 by disjoint intervals. Since these $U_2(U_1, x_2)$ are elements of Markov partitions $\mathcal{R}_2^j$, $j \geq 1$, two of them either are disjoint or have one contained in the other. Thus, we can extract a finite subcovering by disjoint elements. We now define $\mathcal{U}_m$ to be the family of sets $U_1 \times U_2(U_1, x_2)$ obtained in this way for all $U_1 \in \mathcal{R}_1^m$ contained in R_1. This is a covering of $R_1 \times R_2$ by disjoint cubes. Moreover, it is not difficult to deduce from the bounded distortion property that there is $0 < b < 1$ (depending only on K_2 and Ψ_2) such that, denoting by $U_2'(U_1, x_2)$ the element of $\mathcal{R}_2^{\overline{m}(U_1, x_2)+1}$ that contains x_2, we have $\ell(U_2'(U_1, x_2)) \geq b\ell(U_2(U_1, x_2))$. By definition of $U_2'(U_1, x_2)$, we also have $\ell(U_2'(U_1, x_2)) \leq \ell(U_1)$. Therefore,

$$\ell(U_1) \geq b \cdot \ell(U_2(U_1, x_2)). \tag{8}$$

Then, by (7) and (8) we get

$$\sum_{\mathcal{U}_m}(\mathrm{diam}(U_1 \times U_2(U_1, x_2)))^d = \sum_{\mathcal{U}_m}\ell(U_2(U_1, x_2))^d$$

$$\leq b^{-d_1}\sum_{\mathcal{U}_m}\ell(U_1)^{d_1}\ell(U_2(U_1, x_2))^{d_2}$$

$$\leq b^{-d_1}\sum_{\mathcal{U}_m}(a_2' m_{d_1}(U_1 \cap K_1))(a_2'' m_{d_2}(U_2(U_1, x_2)) \cap K_2))$$

$$= b^{-d_1}a_2'a_2''\mu(A).$$

Since the diameter of $\mathcal{U}_m$ may be taken arbitrarily small (by taking m large), we have $m_d(A) \leq b^{-d_1}a_2'a_2''\mu(A)$ and so our argument is complete. □

Note that $0 < m_d(K) < \infty$ ((a) in the proposition) is related to the fact that since K_1 and K_2 are dynamically defined, $HD(K_1 \times K_2) = HD(K_1) + HD(K_2)$. This also follows from the previous theorem stating that $d(K_i) = HD(K_i)$, $i = 1, 2$, and the general product formulas $HD(K_1 \times K_2) \geq HD(K_1) + HD(K_2)$ and $d(K_1 \times K_2) \leq d(K_1) + d(K_2)$ together with the inequality $d(K) \geq HD(K)$.

We now establish an interesting relation between Hausdorff dimension and thickness for Cantor sets in the line. In particular, if the thickness is large then the Hausdorff dimension is close to 1.

PROPOSITION 5. *If $K \subset \mathbb{R}$ is a Cantor with thickness τ then $HD(K) \geq (\log 2/\log(2 + 1/\tau))$.*

PROOF: Let $\beta = (\log 2/\log(2 + 1/\tau))$. We show that $H_\beta(\mathcal{U}) \geq (\mathrm{diam}\, K)^\beta$ for every finite open covering $\mathcal{U}$ of K, which clearly implies the proposition. The key ingredient in this proof is the following elementary fact:

$$\min\{x^\beta + z^\beta \mid x \geq 0, z \geq 0, x + z \leq 1, x \geq \tau(1 - x - z), z \geq \tau(1 - x - z)\} = 1. \tag{9}$$

We assume from now on that $\mathcal{U}$ is a covering with disjoint intervals. This is no restriction because whenever two elements of $\mathcal{U}$ have nonempty intersection we can replace them by their union, getting in this way a new covering $\mathcal{V}$ such that $H_\beta(\mathcal{V}) \leq H_\beta(\mathcal{U})$. Note that, since $\mathcal{U}$ is an open covering of K, it covers all but a finite number of gaps of K. Let U, a gap of K, have minimal length among the gaps of K which are not covered by $\mathcal{U}$. Let C^ℓ and C^r be the bridges of K at the boundary points of U. (See Figure 4.8).

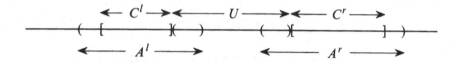

Figure 4.8

By construction there are A^ℓ, $A^r \in \mathcal{U}$ such that $C^\ell \subset A^\ell$ and $C^r \subset A^r$. Take the convex hull A of $A^\ell \cup A^r$. Then

$$\ell(A^\ell) \geq \ell(C^\ell) \geq \tau \cdot \ell(U) \geq \tau(\ell(A) - \ell(A^\ell) - \ell(A^r))$$

and

$$\ell(A^r) \geq \ell(C^r) \geq \tau \cdot \ell(U) \geq \tau(\ell(A) - \ell(A^\ell) - \ell(A^r))$$

and so, by (9), $(\ell(A^\ell))^\beta + (\ell(A^r))^\beta \geq (\ell(A))^\beta$. This means that the covering $\mathcal{U}_1$ of K obtained by replacing A^ℓ and A^r by A in $\mathcal{U}$ is such that $H_\beta(\mathcal{U}_1) \leq H_\beta(\mathcal{U})$. Repeating the argument we eventually construct $\mathcal{U}_k$, a covering of the convex hull of K with $H_\beta(\mathcal{U}_k) \leq H_\beta(\mathcal{U})$. Since we must have $H_\beta(\mathcal{U}_k) \geq (\dim K)^\beta$, this ends the proof.   □

Note that in general there can be no nontrivial upper estimates for the Hausdorff dimension in terms of the thickness, even in the dynamically defined case. To see this, recall the earlier example of an affine Cantor set K with Markov partition $\{K_1, \ldots, K_k\}$ with components all of length $\beta \cdot \operatorname{diam} K$, $0 < \beta < \frac{1}{k}$, and gaps between K_i and K_{i+1} all of length $(1 - \beta \cdot k) \cdot \operatorname{diam}(K)/(k-1)$. As we saw, we have $HD(K) = \log k / \log \beta^{-1}$. The thickness can easily be shown to be $\tau(K) = \beta(k-1)/(1 - \beta \cdot k)$. Now consider a sequence of such Cantor sets characterized by k and β_k such that $\lim_{k \to \infty} k \cdot \beta_k = \alpha \in (0, 1)$. Then, as $k \to \infty$, the Hausdorff dimension tends to 1 while the thickness converges to $\alpha/(1 - \alpha)$.

This fact is not really surprising since the thickness was defined as an infimum and so having $\tau(K)$ small gives very little information concerning the Cantor set. As mentioned before, this was our main motivation for introducing a variation of the thickness which we called denseness. We shall prove that Cantor sets with small denseness have small Hausdorff dimension.

Let us first observe that if K is an affine Cantor set as above and $2\ell - 1 \leq k \leq 2\ell$, then $\theta(K) = (\ell - 1) + \ell\beta(k-1)/(1 - \beta k)$. This follows from the fact that the infimun in the definition of $\theta(K)$ is attained on the presentations $\mathcal{U} = \{U_n\}$ of K satisfying

- if $\ell(U_m) > \ell(U_n)$ then $m < n$,

- for U_m, U_n with $\ell(U_m) = \ell(U_n)$, we consider the maximal interval C, if it exists, containing U_m and U_n but containing no points of gaps longer than U_m or U_n (if dist $(U_m, \partial C) >$ dist $(U_n, \partial C)$ then we require $m < n$).

In particular, for such Cantor sets the thickness and the denseness coincide only when $k = 2$.

PROPOSITION 6. *Let* $K \subset \mathbb{R}$ *be a Cantor set with denseness* θ. *Then* $HD(K) \leq \log 2/\log(2 + 1/\theta)$.

PROOF: Given $\theta_1 > \theta$, let $\mathcal{U} = (U_n)$ be a presentation of K such that $\sup_u \tau(K, \mathcal{U}, u) \leq \theta_1$. For $n \geq 1$, let $\mathcal{A}_n$ be the covering of K formed by connected components of $I - (U_1 \cup U_2 \cup \cdots \cup U_n)$, where I is the minimal interval containing K. We claim that for $\beta = \log 2/\log(2 + 1/\theta_1)$ the sequence $(H_\beta(\mathcal{A}_n))_n$ is bounded. To show this, we first observe that the only difference between consecutive coverings $\mathcal{A}_{n-1}$ and $\mathcal{A}_n$ is that the interval $A_{n-1} \in \mathcal{A}_{n-1}$ containing U_n is replaced by two new intervals $C_n^\ell, C_n^r \in \mathcal{A}_n$. (See Figure 4.9).

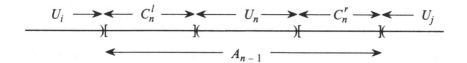

Figure 4.9

Then we have $H_\beta(\mathcal{A}_n) - H_\beta(\mathcal{A}_{n-1}) = [\ell(C_n^\ell)]^\beta + [\ell(C_n^r)]^\beta - [\ell(A_{n-1})]^\beta$. On the other hand, the assumptions in the proposition imply $\ell(C_n^\ell) \leq \theta_1 \cdot \ell(U_n)$ and $\ell(C_n^r) \leq \theta_1 \cdot \ell(U_n)$. Now using

$$\max\{x^\beta + z^\beta \mid x \geq 0, z \geq 0, x+z \leq 1, x \leq \theta_1 \cdot (1-x-z), z \leq \theta_1 \cdot (1-x-z)\} = 1,$$

we get $[\ell(C_n^\ell)]^\beta + [\ell(C_n^r)]^\beta \leq [\ell(A_{n-1})]^\beta$, that is $H_\beta(\mathcal{A}_n) \leq H_\beta(\mathcal{A}_{n-1})$. Therefore, the sequence $(H_\beta(\mathcal{A}_n))_n$ is nonincreasing and so it is bounded as claimed. Since the diameters of $\mathcal{A}_n$ clearly converge to zero this implies $m_\beta(K) < \infty$ and so $HD(K) \leq \beta = \log 2 / \log(2 + 1/\theta_1)$. Since $\theta_1 > \theta$ is arbitrary, the proposition follows. □

PROPOSITION 7. *If K is a dynamically defined Cantor set then $0 < \tau(K) \leq \theta(K) < \infty$ and so $0 < d(K) = HD(K) < 1$.*

PROOF: We have already seen that the denseness is always larger than or equal to the thickness. Let us show that for some presentation $\mathcal{U}$ of K

$$0 < \inf_u \tau(K, \mathcal{U}, u) \leq \sup_u \tau(K, \mathcal{U}, u) < \infty. \tag{10}$$

This immediately implies the first statement in the proposition. The second statement is a direct consequence of the first one and the two last propositions above. To construct $\mathcal{U}$ we proceed as follows. Let $\mathcal{R} = \{K_1, \ldots, K_k\}$ be a Markov partition for K and $\tilde{U}_1, \ldots, \tilde{U}_{k-1}$ be the gaps of K between the intervals in this partition. For any gap U of K let $s(U) \geq 0$ be the smallest integer such that $\Psi^{s(U)}(U)$ is not contained in any $K_i \in \mathcal{R}$ ($s(U) = 0$ if and only if $U \in \{\tilde{U}_1, \ldots, \tilde{U}_{k-1}\}$ or U is unbounded). Clearly, given any $\bar{s} \geq 0$ the set of gaps U of K such that $s(U) \leq \bar{s}$ is finite. Therefore we may take $\mathcal{U} = \{U_n\}$ an ordering of the bounded gaps of K such that

$$i \leq j \Rightarrow s(U_i) \leq s(U_j).$$

We now prove that such $\mathcal{U}$ satisfies (10). Let $u \in \partial U_i$ and C be the $\mathcal{U}$-component of K at u. Observe that $\Psi^n(C)$ is contained in some $K_{i_n} \in \mathcal{R}$, for all $0 \leq n \leq s(U_i) - 1$. Otherwise C would contain a gap U_j with $s(U_j) < s(U_i)$ and so $j < i$, which contradicts the definition of $\mathcal{U}$-component. Then by the bounded distortion property there is $a > 0$, depending only on (K, Ψ), such that

$$a^{-1} \cdot \frac{\ell(C)}{\ell(U_i)} \leq \frac{\ell(\Psi^{s(U_i)}(C))}{\ell(\Psi^{s(U_i)}(U_i))} \leq a \cdot \frac{\ell(C)}{\ell(U_i)},$$

i.e.

$$a^{-1} \cdot \frac{\ell(\Psi^{s(U_i)}(C))}{\ell(\Psi^{s(U_i)}(U_i))} \leq \tau(K, \mathcal{U}, u) \leq a \cdot \frac{\ell(\Psi^{s(U_i)}(C))}{\ell(\Psi^{s(U_i)}(U_i))}.$$

To complete the proof it is now sufficient to show that the values of $\frac{\ell(\Psi^{s(U_i)}(C))}{\ell(\Psi^{s(U_i)}(U_i))}$ that we obtain in this way can be bounded from zero and infinity. To see this observe first that we must have $\Psi^{s(U_i)}(U_i) \in \{\tilde{U}_1, \ldots, \tilde{U}_{k-1}\}$ and so $\ell(\Psi^{s(U_i)}(U_i))$ can take only a finite set of values. As to $\Psi^{s(U_i)}(C)$, note that its length cannot exceed $\operatorname{diam}(K)$ and that, on the other hand, it must contain some $K_r \in \mathcal{R}$. This last affirmative is proved as follows. Let v be the other boundary point of C and U the gap of K such that $v \in \partial U$. Then either U is unbounded or $U = U_j$ for some $j < i$. In either case we have $s(U) \le s(U_i)$. It follows that $\Psi^{s(U_i)}(v)$ must be in the boundary of some K_r, which then must be contained in $\Psi^{s(U_i)}(C)$. We conclude that $\ell(\Psi^{s(U_i)}(C))$ is also bounded away from zero and infinity. This completes the proof of the proposition. □

It follows easily from the definition of Hausdorff dimension that, for any subset K of $\mathbb{R}$, we have

$$HD(K) < 1 \Longrightarrow m(K) = 0,$$

where m denotes the Lebesgue measure. *As an important particular case, the dynamically defined Cantor sets have always Lebesgue measure zero.* This conclusion remains valid for Cantor sets defined by two–dimensional diffeomorphisms which are *only of class* C^2, since, as observed in the last remark of Section 1, such Cantor sets still have the bounded distortion property.

The situation is rather different for diffeomorphisms which are only once differentiable. We describe briefly a construction, due to Bowen [**B**,1975b], of a C^1 diffeomorphism with an invariant horseshoe $\Lambda = \overline{W^s(p) \cap W^u(p)}$, p a hyperbolic fixed point, such that $m(\Lambda) > 0$ and $m(W^s(p) \cap \Lambda) > 0$. In this particular example, $W^s(p) \cap \Lambda$ is invariant under an expanding C^1 map but the bounded distortion property no longer holds.

Given a sequence $\{\beta_n\}$ of positive real numbers satisfying

$$\sum_{n \ge 0} \beta_n < 2 \quad \text{and} \quad \frac{\beta_n}{\beta_{n+1}} \xrightarrow[n \to \infty]{} 1,$$

we construct in $J = [-1, 1]$, by the standard procedure, a Cantor set K_J in such way that at the nth step we remove 2^n intervals, $J_{n,k}$, $k \in \{1, \ldots, 2^n\}$, of length $\frac{\beta_n}{2^n}$. It is then clear that $m(K_J) = 2 - \sum_{n \ge 0} \beta_n$ is positive. For each $n \ge 1$ and $k \in \{2^{n-1} + 1, \ldots, 2^n\}$, define

$$g_{n,k} \colon J_{n,k} \longrightarrow J_{n-1,k-2^{n-1}}$$

as follows:

(i) $g_{n,k}$ is a C^1 orientation preserving homeomorphism;

(ii) $g'_{n,k}(a_{n,k}) = g'_{n,k}(b_{n,k}) = 2$, where $J_{n,k} = [a_{n,k}, b_{n,k}]$;

(iii) $\displaystyle\sup_{x\in J_{n,k}} |2 - g'_{n,k}(x)| \xrightarrow[n\to+\infty]{} 0.$

The choice of the β_n's guarantees that (i)–(iii) coexist and makes possible this construction.

Now, from the above conditions, we can continuously extend all the $g_{n,k}$'s to a homeomorphism g of class C^1 on $[\beta_0/2, 1]$, so that $g'|_{K_J} \equiv 2$.

Finally, let Q be $J \times J$ and $\varphi: Q \to Q$ be a diffeomorphism given by

$$\varphi(x,y) = (g(x), g^{-1}(y)), \qquad \text{if}\quad x \in \left[\frac{\beta_0}{2}, 1\right],$$

$$\varphi(x,y) \notin Q, \qquad \text{if}\quad |x| < \frac{\beta_0}{2},$$

$$\varphi(x,y) = (g(-x), -g^{-1}(y)), \quad \text{if}\quad x \in \left[-1, -\frac{\beta_0}{2}\right].$$

(See Figure 4.10)

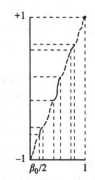

graph of g, partially defined

Figure 4.10

The reader may easily verify that φ is of class C^1, $p = (1, -1)$ is a hyperbolic fixed point of φ, $\Lambda = \displaystyle\bigcap_{n\in\mathbf{Z}} \varphi^n(Q) = K_J \times K_J$ is a hyperbolic horseshoe and that $W^s_{\mathrm{loc}}(p) \cap \Lambda = K_J$. Besides, both Λ and K_J have Lebesgue measure greater than zero.

§3   Local invariants and continuity

We conclude this chapter with some relevant facts on localized versions of the numerical invariants for Cantor sets introduced so far, and on the (continuous) dependence of these invariants on the Cantor set, at least for dynamically defined Cantor sets.

We give the definition of local thickness; local denseness, local Hausdorff dimension and local limit capacity are similarly defined. Let $K \subset \mathbb{R}$ be a Cantor set and $k \in K$. The local thickness $\tau_{\mathrm{loc}}(K, k)$ of K at k is defined as

$$\tau_{\mathrm{loc}}(K, k) = \lim_{\varepsilon \to 0} \sup\{\tau(\tilde{K}) \mid \tilde{K} \text{ is the intersection of } K \text{ with an interval}$$
$$\text{contained in an } \varepsilon\text{-neighbourhood of } k\}.$$

For dynamically defined Cantor sets these notions have some additional properties. Let K be a dynamically defined Cantor set with expanding map Ψ. Then for every $U \subset K$, U open, there is some n so that $\Psi^n(U) = K$. From this and the bounded distortion property it follows that the local invariants $\tau_{\mathrm{loc}}(K, k)$, $\theta_{\mathrm{loc}}(K, k)$, $HD_{\mathrm{loc}}(K, k)$, and $d_{\mathrm{loc}}(K, k)$ are, in the dynamically defined case, all independent of k. Also, since the limit capacity and the Hausdorff dimension are invariant under diffeomorphisms, one has in this case $HD_{\mathrm{loc}}(K, k) = HD(K) = d(K) = d_{\mathrm{loc}}(K, k)$. The thickness and the denseness are not invariant under diffeomorphisms, and we may have $\tau(K) < \tau_{\mathrm{loc}}(K, k)$ or $\theta(K) < \theta_{\mathrm{loc}}(K, k)$.

For a discussion of the continuous dependence of the invariants on the Cantor set, we restrict ourselves to the dynamically defined case. *Bearing in mind the dynamics of basic sets of surface diffeomorphisms, we define when two Cantor sets are near each other as follows.* Let K be a Cantor set with expanding map Ψ and Markov partition $\{K_1, \ldots, K_\ell\}$. Suppose that Ψ is $C^{1+\varepsilon}$ with Hölder constant C, i.e. with $|\Psi'(p) - \Psi'(q)| \leq C |p - q|^\varepsilon$ for all p, q in a neighbourhood of K. We say that the Cantor set $\tilde{K}$ is near K if $\tilde{K}$ has expanding map $\tilde{\Psi}$ and Markov partition $\tilde{K}_1, \ldots, \tilde{K}_\ell$ such that

- $\tilde{\Psi}$ is $C^{1+\tilde{\varepsilon}}$ and is C^1 near Ψ, its derivative $\tilde{\Psi}'$ has Hölder constant $\tilde{C}$ such that $(\tilde{\varepsilon}, \tilde{C})$ is near (ε, C),

- $(\tilde{K}_1, \ldots, \tilde{K}_\ell)$ is near $(K_1, \ldots, K_\ell)$ in the sense that corresponding end-points are near.

An important consequence of this definition is the existence, for nearby Cantor sets K and $\tilde{K}$ as above, of a homeomorphism $h : K \to \tilde{K}$, C^0-close to the identity, such that $\tilde{\Psi} \circ h = h \circ \Psi$. We construct h as follows. Notice first that, because of the proximity assumptions in the definition, $\Psi(K_i)$ intersects (and then contains) K_j if and only if the same happens with $\tilde{\Psi}(\tilde{K}_i)$ and $\tilde{K}_j$. It follows that, given $x \in K$, there is $\tilde{x} \in \tilde{K}$ such that $\tilde{\Psi}^n(\tilde{x}) \in \tilde{K}_i \Leftrightarrow \Psi^n(x) \in K_i$, for all $n \geq 0$. Since $\tilde{\Psi}$ is expanding, $\tilde{x}$ must be unique; we define $h(x) = \tilde{x}$. Clearly $\tilde{\Psi}(h(x)) = h(\Psi(x))$. On the other hand we can obtain h^{-1} by a symmetrical construction, so h is really a bijection. Checking that h is close to the identity presents no particular difficulty. Just construct Markov partitions $\mathcal{R}^n$ and $\tilde{\mathcal{R}}^n$ for K and $\tilde{K}$ as in the previous section taking connected components of the inverse images of the K_j, respectively $\tilde{K}_j$, by Ψ^{n-1}, respectively $\tilde{\Psi}^{n-1}$. Then, one observes that x and $h(x)$ belong to corresponding intervals of $\mathcal{R}^n$ and $\tilde{\mathcal{R}}^n$ for all n and that corresponding intervals are uniformly (meaning independently of n) close, due to the closeness of $\tilde{K}$ to K, Ψ to $\tilde{\Psi}$ and to the bounded distortion property. We are left to show that h is continuous. We do more than that: we prove that it is Hölder continuous. Take $\delta > 0$ such that $d(K_i, K_j) > 3\delta$ and $d(\tilde{K}_i, \tilde{K}_j) > 3\delta$ for all $i \neq j$. Now, for $x, y \in K$ with $|x - y| \leq \delta$ we let $n = n(x,y) \geq 0$ be such that

$$\left|\Psi^i(x) - \Psi^i(y)\right| \leq 2\delta \text{ for } 0 \leq i \leq n - 1$$

and

$$|\Psi^n(x) - \Psi^n(y)| \geq 2\delta.$$

By the definition of δ, the interval $[\Psi^i(x), \Psi^i(y)]$ is contained in some element of the Markov partition, for every $0 \leq i \leq n-1$. On the other hand we may assume that

$$\left|\tilde{\Psi}^i(\tilde{x}) - \tilde{\Psi}^i(\tilde{y})\right| \leq 3\delta \text{ for } 0 \leq i \leq n - 1.$$

To have this we just take $\tilde{K}$ close enough to K, in order to have $|h(x) - x| \leq \delta/2$ for all x (note that $\tilde{\Psi}^i(\tilde{x}) = h(\Psi^i(x))$). Then again $[\tilde{\Psi}^i(\tilde{x}), \tilde{\Psi}^i(\tilde{y})]$ must be contained in some $\tilde{K}_j$, for all $0 \leq i \leq n - 1$. By the mean value theorem there are $\xi_i \in [\Psi^i(x), \Psi^i(y)]$, $\tilde{\xi}_i \in [\tilde{\Psi}^i(\tilde{x}), \tilde{\Psi}^i(\tilde{y})]$ such that

$$|\Psi^n(x) - \Psi^n(y)| = |x - y| \prod_0^{n-1} |\Psi'(\xi_i)|,$$

$$\left|\tilde{\Psi}^n(\tilde{x}) - \tilde{\Psi}^n(\tilde{y})\right| = |\tilde{x} - \tilde{y}| \prod_0^{n-1} \left|\tilde{\Psi}'(\tilde{\xi}_i)\right|.$$

Clearly, we can take $0 < \gamma < 1$ such that $|\Psi'(\xi_i)|^\gamma \leq \left|\tilde{\Psi}'(\tilde{\xi}_i)\right|$. Then we get

$$\frac{|\tilde{x} - \tilde{y}|}{|x - y|^\gamma} \leq \frac{\left|\tilde{\Psi}^n(\tilde{x}) - \tilde{\Psi}^n(\tilde{y})\right|}{|\Psi^n(x) - \Psi^n(y)|^\gamma} \leq (\text{ diam } K) \cdot \delta^{-\gamma}.$$

Hence h is Hölder continuous, as we claimed.

In fact we have proven even more. Since $\tilde{\Psi}$ is C^1-close to Ψ and $\tilde{\xi}_i$ is close to ξ_i, the values of $|\Psi'(\xi_i)|$ and $\left|\tilde{\Psi}'(\tilde{\xi}_i)\right|$ are almost equal. Therefore, if $\tilde{K}$ is close to K, *the Hölder exponent γ of the conjugacy h may be taken close to 1.* This and the analogous fact for h^{-1} imply the following important result ([**MM**,1983], [**PV**,1988]; see Remark 1 at the end of the section).

THEOREM 1. *The Hausdorff dimension and limit capacity of a dynamically defined Cantor set K depend continuously on K.*

To derive the theorem from the considerations above one just has to observe that the existence of a homeomorphism $h: K \to \tilde{K}$ such that h and h^{-1} both are C^γ implies that $\gamma \cdot HD(K) \leq HD(\tilde{K}) \leq \gamma^{-1} \cdot HD(K)$ (and analogously for limit capacity). This, in its turn, is a direct consequence of the definitions. This proof is similar to that in [**PV**,1988].

Now we state and prove the corresponding result for thickness and denseness.

THEOREM 2. *The thickness and the denseness of a dynamically defined Cantor set K depend continuously on K. The same holds for local thickness and local denseness.*

Heuristically, the theorem is proved as follows. The global strategy is to show that the values $\tau(K, \mathcal{U}, u)$, with $\mathcal{U} = \{U_n\}$ a presentation of K and u in the boundary of some bounded gap $U = U_n$, depend equicontinuously on K in the sense that if $\tilde{K}$ is close to K then $\tau(\tilde{K}, h(\mathcal{U}), h(u))$ is close to $\tau(K, \mathcal{U}, u)$ for all $\mathcal{U}$ and u. Here $h: K \to \tilde{K}$ is the conjugacy from Ψ to $\tilde{\Psi}$ described above (assume $\tilde{K}$ close enough to K to ensure that h exists) and $h(\mathcal{U})$ is the presentation of $\tilde{K}$ given by $h(\mathcal{U}) = \{h(U_n)\}$, where $h(U_n)$ is defined by $\partial h(U_n) = h(\partial U_n)$. Observe that h, as we constructed it, is monotonic.

For any given u and $\mathcal{U}$ we can, just by forcing h to be close enough to the identity, make $\tau(\tilde{K}, h(\mathcal{U}), h(u))$ arbitrarily close to $\tau(K, \mathcal{U}, u)$. We can even make this happen simultaneously for all u (and $\mathcal{U}$) for which the corresponding gap U is big, say with length bigger than some fixed $\alpha > 0$. However, such a simple argument is insufficient to obtain the uniform closeness that we need. To deal with the small gaps we must use the bounded distortion property. The idea is to iterate the gap U and the $\mathcal{U}$-component C of u until they become big. To be precise we fix $\beta > 0$ and take $k = k(U, C) \geq 0$ minimal such that $\ell(\Psi^k(U \cup C)) \geq \beta$. From the bounded distortion property we conclude that $\tau(K, \mathcal{U}, u) = \dfrac{\ell(C)}{\ell(U)}$ is almost equal to $\dfrac{\ell(\Psi^n(C))}{\ell(\Psi^n(U))}$: their ratio admits a bound depending only on $\beta > 0$ and K and which can be made arbitrarily close to 1 by taking β small enough. Analogously, from the bounded distortion property for $\tilde{K}, \tilde{\Psi}$ we obtain that $\tau(\tilde{K}, h(\mathcal{U}), h(u)) = \dfrac{\ell(h(C))}{\ell(h(U))}$ is almost equal to $\dfrac{\ell(\tilde{\Psi}^n(h(C)))}{\ell(\tilde{\Psi}^n(h(U)))}$. Moreover, and this is a key point, the bound for the ratio of these last two values may be taken to be independent of $\tilde{K}$ in a neighbourhood of K. This is a consequence of the fact that bounds for the distortion may be taken to be uniform in a neighbourhood of any Cantor set. To explain this, let us first observe that the positive numbers $c(\delta)$ constructed in the proof of the bounded distortion property vary continuously with the dynamically defined Cantor set. In fact, these $c(\delta)$ depend only on the Hölder constants of the derivative of the expanding map and, by definition, nearby Cantor sets have expanding maps whose derivatives have nearby Hölder constants. In particular, it follows that we can take (new) upper bounds $c(\delta)$ as in the statement of the bounded distortion property which are uniform, i.e. independent of the Cantor set in a neighbourhood of K. We assume in what follows that $\tilde{K}$ belongs to this neighbourhood.

Now, if $\ell(\Psi^k(U))$ is big, that is larger than α, we can argue as before, i.e. use the proximity of h to the identity to conclude that $\dfrac{\ell(\tilde{\Psi}^k(h(C)))}{\ell(\tilde{\Psi}^k(h(C)))} = \dfrac{\ell(h(\Psi^k(C)))}{\ell(h(\Psi^k(U)))}$ is close to $\dfrac{\ell(\Psi^k(C))}{\ell(\Psi^k(U))}$. This, together with the estimates obtained above with the aid of the bounded distortion property, proves that $\tau(\tilde{K}, h(\mathcal{U}), h(u))$ is close to $\tau(K, \mathcal{U}, u)$, as we wanted to show.

Of course, we still have the problem that $\Psi^k(U)$ may be small. Iterating further is no solution: it may not be possible to do it, if $\Psi^j(C)$ gets out of the domain of Ψ before $\Psi^j(U)$ gets large. Even if this does not happen, as we iterate the length of $\Psi^j(U \cup C)$ gets bigger and so the bounds given by the bounded distortion property get rougher. Clearly, for the preceding argument we needed these bounds to be close to 1. Instead, what we do is to show that for our purposes this situation doesn't need to be taken into consideration. First, we observe that since $\ell(\Psi^k(U)) \leq \alpha$ and $\ell(\Psi^k(U \cup C)) \geq \beta$, if we have chosen from the beginning $\beta \gg \alpha$, then $\dfrac{\ell(\Psi^k(C))}{\ell(\Psi^k(U))}$ must be very big. The conjugacy h being close to the identity, the same holds for $\dfrac{\ell(\tilde{\Psi}^k(h(C)))}{\ell(\tilde{\Psi}^k(h(U)))}$. Using the bounded distortion property as above we conclude that $\tau(K, \mathcal{U}, u)$ and $\tau(\tilde{K}, h(\mathcal{U}), h(u))$ are very big. Since in the calculation of both the thickness and the denseness one must at some point take an infimum, these values are irrelevant for this calculation and so may be disregarded when proving the continuity of $\theta(K)$ and $\tau(K)$.

We now come to a formal proof.

PROOF: Let $A = \sup |\Psi'|$ and $B = 2\theta(K) + 8$. Let $\varepsilon > 0$, $\delta > 0$ and $\alpha > 0$. Suppose that $\tilde{K}$ is close enough to K so that $|h(x) - x| \leq \alpha\delta$ for all $x \in K$. We prove that if $\alpha > 0$ and $\delta > 0$ are chosen appropriately small (the precise conditions are given below) then this implies

(a) $\qquad\qquad \theta(\tilde{K}) \leq (1+\varepsilon)^2 \theta(K) + \varepsilon(1+\varepsilon),$

(b) $\qquad\qquad \theta(\tilde{K}) \geq (1+\varepsilon)^{-2} \theta(K) - \varepsilon(1+\varepsilon)^{-1},$

(c) $\qquad\qquad \tau(\tilde{K}) \leq (1+\varepsilon)^2 \tau(K) + \varepsilon(1+\varepsilon),$

(d) $\qquad\qquad \tau(\tilde{K}) \geq (1+\varepsilon)^{-2} \tau(K) - \varepsilon(1+\varepsilon)^{-1}.$

This proves the first part of the theorem. Then we show that the second part is an easy consequence of the first one.

First we take $\alpha > 0$ small enough so that the $2AB\alpha$-neighbourhood of K is contained in the domain of Ψ. Clearly, we may assume that the same holds for $\tilde{K}$ and $\tilde{\Psi}$. For $\mathcal{U} = \{U_n\}$ a presentation of K, u a boundary point of a bounded gap $U = U_n$ and C the $\mathcal{U}$ component of K at u, take $k \geq 0$ minimal such that $\ell(\Psi^k(U \cup C)) \geq B\alpha$. Then $\ell(\Psi^k(U \cup C)) \leq AB\alpha$ (because $\ell(\Psi^{k-1}(U \cup C)) \leq B\alpha$) and so $\ell(\tilde{\Psi}^k(h(U) \cup h(C))) \leq AB\alpha + 2\alpha\delta \leq 2AB\alpha$ (as long as $\delta \leq \frac{AB}{2}$). By the bounded distortion property we have

$$e^{-c(AB\alpha)} \leq \left[\left(\frac{\ell(\Psi^k(C))}{\ell(\Psi^k(U))}\right) \Big/ \left(\frac{\ell(C)}{\ell(U)}\right)\right] \leq e^{c(AB\alpha)} \tag{1}$$

and

$$e^{-c(2AB\alpha)} \leq \left[\left(\frac{\ell(\tilde{\Psi}^k(h(C)))}{\ell(\tilde{\Psi}^k(h(U)))}\right) \Big/ \frac{\ell(h(C))}{\ell(h(U))}\right] \leq e^{c(2AB\alpha)} \tag{2}$$

where $c(\ )$ is the distortion bounding function that we recalled in the heuristic proof. We assume that α is small enough so that this implies

$$(1+\varepsilon)^{-1} \leq \left[\left(\frac{\ell(\Psi^k(C))}{\ell(\Psi^k(U))}\right) \Big/ \left(\frac{\ell(C)}{\ell(U)}\right)\right] \leq (1+\varepsilon), \tag{1a}$$

$$(1+\varepsilon)^{-1} \leq \left[\left(\frac{\ell(\tilde{\Psi}^k(h(C)))}{\ell(\tilde{\Psi}^k(h(U)))}\right) \Big/ \left(\frac{\ell(h(C))}{\ell(h(U))}\right)\right] \leq (1+\varepsilon). \tag{2a}$$

Now we distinguish two cases according to the size of Ψ. Suppose first that $\ell(\Psi^k(U)) \geq \alpha$. Then

$$\left|\frac{\ell(\Psi^k(C))}{\ell(\Psi^k(U))} - \frac{\ell(h(\Psi^k(C)))}{\ell(h(\Psi^k(U)))}\right|$$

$$\leq \frac{\ell(\Psi^k(C)) \cdot \left|\ell(h(\Psi^k(U))) - \ell(\Psi^k(U))\right| + \ell(\Psi^k(U)) \cdot \left|\ell(h(\Psi^k(C))) - \ell(\Psi^k(C))\right|}{\ell(\Psi^k(U)) \cdot \ell(h(\Psi^k(U)))}$$

$$\leq \frac{AB\alpha \cdot 2\alpha\delta + AB\alpha \cdot 2\alpha\delta}{\alpha \cdot (\alpha - 2\alpha\delta)} = \delta \frac{4AB}{1 - 2\delta} \cdot \tag{3}$$

If $\delta > 0$ is sufficiently small this implies

$$\left| \frac{\ell(\Psi^k(C))}{\ell(\Psi^k(U))} - \frac{\ell(\tilde{\Psi}^k(h(C)))}{\ell(\tilde{\Psi}^k(h(U)))} \right| \le \varepsilon. \tag{3a}$$

From (1a), (2a) and (3a) it immediately follows that

$$\tau(\tilde{K}, h(\mathcal{U}), h(u)) \le (1 + \varepsilon) \cdot ((1 + \varepsilon) \cdot \tau(K, \mathcal{U}, u) + \varepsilon) \tag{4a}$$

and

$$\tau(\tilde{K}, h(\mathcal{U}), h(u)) \ge (1 + \varepsilon)^{-1}((1 + \varepsilon)^{-1}\tau(K, \mathcal{U}, u) - \varepsilon). \tag{4b}$$

Let now $\ell(\Psi^k(U)) \le \alpha$. Then we must have $\ell(\Psi^k(C)) \ge B\alpha - \alpha$. Moreover $\ell(\tilde{\Psi}^k(h(U))) \le \alpha + 2\alpha\delta$ and $\ell(\tilde{\Psi}^k(h(C))) \ge B\alpha - \alpha - 2\alpha\delta$. This together with (1) and (2) implies

$$\frac{\ell(C)}{\ell(U)} \ge e^{-c(AB\alpha)} \cdot \frac{B\alpha - \alpha}{\alpha} = (B - 1)e^{-c(AB\alpha)} \tag{5}$$

and

$$\frac{\ell(h(C))}{\ell(h(U))} \ge e^{-c(2AB\alpha)} \cdot \frac{B\alpha - 2\alpha\delta - \alpha}{\alpha + 2\alpha\delta} = \frac{B - 2\delta - 1}{1 + 2\delta} \cdot e^{-c(2AB\alpha)}. \tag{6}$$

Since we have chosen $B = 2\theta(K) + 8$, we can suppose α and δ small enough so that these relations imply

$$\tau(K, \mathcal{U}, u) \ge (\theta(K) + 3) \tag{5a}$$

and

$$\tau(\tilde{K}, h(\mathcal{U}), h(u)) \ge (\theta(K) + 3). \tag{6a}$$

Now we proceed to prove the affirmatives (a) through (d) stated near the beginning of the proof. Recall that by definition

$$\tau(K) = \sup_{\mathcal{U}} \inf_{u} \tau(K, \mathcal{U}, u),$$

$$\theta(K) = \inf_{\mathcal{U}} \sup_{u} \tau(K, \mathcal{U}, u).$$

To prove (a) we must find for any given $\mathcal{U}$ a presentation $\tilde{\mathcal{U}}$ of $\tilde{K}$ such that

$$\sup_{\tilde{u}} \tau(\tilde{K}, \tilde{\mathcal{U}}, \tilde{u}) \le (1 + \varepsilon)^2 \sup_{u} \tau(K, \mathcal{U}, u) + \varepsilon(1 + \varepsilon) \tag{a1.}$$

There is no loss of generality if we assume that

$$\sup_{u} \tau(K, \mathcal{U}, u) \le \theta(K) + 1. \tag{a2}$$

Take $\tilde{\mathcal{U}} = h(\mathcal{U})$. From (a2) it follows that $\tau(K,\mathcal{U},u) \leq \theta(K) + 1$ for all u and so (5a) never holds. Then, by the previous discussion we must have

$$\tau(\tilde{K},\tilde{\mathcal{U}},\tilde{u} = h(u)) \leq (1 + \varepsilon)^2 \cdot \tau(K,\mathcal{U},u) + \varepsilon(1 + \varepsilon) \tag{4a}$$

(as well as (4b)) for all u. This immediately implies (a1) and so (a) is proved.

The proof of (b) is almost dual to the preceding one so we don't write it down in detail. The only asymmetry comes from the fact that (6a) involves $\theta(K)$ and not $\theta(\tilde{K})$. This is bypassed as follows. First, we may as above suppose that

$$\sup_{\tilde{u}} \tau(\tilde{K},\tilde{\mathcal{U}},\tilde{u}) \leq \theta(\tilde{K}) + 1. \tag{b2}$$

Now, from (a) (which we have already proved) we get that if $\tilde{K}$ is sufficiently near K then $\theta(\tilde{K}) \leq \theta(K) + 1$. Then (b2) implies

$$\sup_{\tilde{u}} \tau(\tilde{K},\tilde{\mathcal{U}},\tilde{u}) \leq \theta(K) + 2 \tag{b3}$$

and now the argument proceeds as before.

To prove (c) we take, for each $\tilde{\mathcal{U}}$, $\mathcal{U} = h^{-1}(\tilde{\mathcal{U}})$ and show that

$$\inf_u \tau(\tilde{K},\tilde{\mathcal{U}},\tilde{u}) \leq (1 + \varepsilon)^2 \inf_u \tau(K,\mathcal{U},u) + \varepsilon(1 + \varepsilon). \tag{c1}$$

To do this we must to each u associate $\tilde{u}$ such that

$$\tau(\tilde{K},\tilde{\mathcal{U}},\tilde{u}) \leq (1 + \varepsilon)^2 \cdot \tau(K,\mathcal{U},u) + \varepsilon(1 + \varepsilon). \tag{c2}$$

Again, it is sufficient to consider the points u for which

$$\tau(K,\mathcal{U},u) \leq \inf_u \tau(K,\mathcal{U},u) + 1. \tag{c3}$$

Take $\tilde{u} = h(u)$ and observe that if (c3) holds then $\tau(K,\mathcal{U},u) \leq \tau(K) + 1 \leq \theta(K) + 1$ and so (5a) doesn't hold. Therefore (4a) is true, and this is just (c2). The proof of (c) is complete.

The proof of (d) is dual to the one of (c) (recall also the remark in the proof of (b)) so we are done with proving the continuity of (global) thickness and denseness.

Finally, recall that the local thickness of a Cantor set K at a point $k \in K$ is defined by

$$\tau_{\text{loc}}(K,k) = \lim_{\delta \to 0}(\sup\{\tau(K_1)|K_1 \subset K \cap B_\delta(k) \text{ a Cantor set }\}).$$

Let $\varepsilon > 0$ be small. Given $\tilde{\delta} > 0$, take $\delta > 0$ such that $h(K \cap B_\delta(k)) \subset \tilde{K} \cap B_{\tilde{\delta}}(h(k))$. Let K_1 be a Cantor set in $K \cap B_\delta(k)$ and let $\tilde{K}_1 = h(K_1)$. If

h is close enough to the identity (i.e. if $\tilde{K}$ is close enough to K) then the arguments above imply $\tau(\tilde{K}_1) \geq \tau(K_1) - \varepsilon$. Since K_1 is arbitrary it follows that

$$\sup\{\tau(K_1) | K_1 \subset K \cap B_\delta(k) \text{ a Cantor set } \}$$
$$\leq \sup\{\tau(\tilde{K}_1) | \tilde{K}_1 \subset \tilde{K} \cap B_{\tilde{\delta}}(h(k)) \text{ a Cantor set } \} + \varepsilon.$$

By making $\tilde{\delta} \to 0$ (and so $\delta \to 0$) we get

$$\tau_{\text{loc}}(K, k) \leq \tau_{\text{loc}}(\tilde{K}, h(k)) + \varepsilon.$$

In the same way one shows

$$\tau_{\text{loc}}(K, k) \geq \tau_{\text{loc}}(\tilde{K}, h(k)) - \varepsilon.$$

This shows the continuity of local thickness. For local denseness the argument is the same. The proof of the theorem is now complete.   □

REMARK 1: Consider a C^3 diffeomorphism φ of a surface, with a basic set Λ and a saddle point $p \in \Lambda$. For $\tilde{\varphi}$ a C^3 nearby diffeomorphism there are $\tilde{\Lambda}$, a basic set, and $\tilde{p} \in \tilde{\Lambda}$, a saddle point (near Λ and p, respectively), and the dynamically defined Cantor sets $W^u(p) \cap \Lambda$ and $W^u(\tilde{p}) \cap \tilde{\Lambda}$ are near in the above sense (if we take nearby parametrizations for $W^u(p)$ and $W^u(\tilde{p})$ as in Section 1 of this present chapter). This follows from the continuous dependence on the diffeomorphism of basic sets and their $C^{1+\varepsilon}$–stable and unstable foliations; see Appendix 1 and Remark 2 in Appendix 2, concerning continuous dependence of Markov partitions. From this and the propositions that we just proved, we deduce the continuous dependence, with respect to the diffeomorphism in the C^3–topology, of all the invariants of $W^u(p) \cap \Lambda$ that we have discussed, namely Hausdorff dimension, limit capacity, thickness and denseness. To show this, one uses the arguments above together with the observation that C^3 diffeomorphisms and C^3 closeness are used only to obtain $C^{1+\varepsilon}$ expanding maps with nearby Hölder constants for the derivatives. This in turn provides bounds for the distortion of distances which are uniform in neighbourhoods of the diffeomorphism and the Cantor set. But, as we remarked before, at the end of Section 1, C^2 diffeomorphisms induce Cantor sets satisfying the bounded distortion property (and the resulting expanding maps are indeed $C^{1+\varepsilon}$ for some $\varepsilon > 0$). The argument that we used there also yields uniform estimates for the distortion in a C^2 neigbourhood of the original diffeomorphism. *Thus, all the above invariants of $W^u(p) \cap \Lambda$ depend continuously on φ in the C^2 topology.*

For Hausdorff dimension and limit capacity, one can go even further: in [**PV**,1988] it is proved that the *Hausdorff dimension and limit capacity of $W^u(p) \cap \Lambda$ depend continuously on the diffeomorphism in the C^1 topology.* This is done by using, as above, conjugacies with Hölder constants near 1. We observe that this result had been obtained in [**MM**,1983] as a consequence of a variational principle of the thermodynamical formalism.

CHAPTER 5

HOMOCLINIC BIFURCATIONS:
FRACTAL DIMENSIONS AND
MEASURE OF BIFURCATION SETS

In this chapter we bring together the Theory of Fractal Dimensions and Bifurcation Theory in Dynamical Systems.

A number of (mostly recent) results, leading to further questions and conjectures as laid out in Chapter 7, shows that the first theory is of fundamental importance to the second, at least in the context of homoclinic bifurcations of nonconservative (say dissipative) systems. The results can be stated in great generality if we focus our attention on the *maximal invariant set* of the restriction of the dynamics to a *neighbourhood of the orbit of homoclinic tangency and an associated basic set*. These results become of a *global nature* when global assumptions are made concerning *filtrations and hyperbolicity of the positive or negative limit set*.

We now explain further the results, but leave the formal statements for the sections following this introduction. So, in this chapter, we will consider one-parameter families φ_μ of *surface* diffeomorphisms which, as the parameter varies, go through a homoclinic tangency say at $\mu = 0$ which we assume to be parabolic and to unfold generically; see Chapter 3. We want to know how *big in the parameter space*, near this bifurcating point, is the *set of values that correspond to diffeomorphisms with a hyperbolic limit set*. A main result here states that this set has a relatively large Lebesgue measure if we assume *small limit capacities* (or *Hausdorff dimensions*) of the stable and unstable sets of Λ, where Λ is the basic set associated with the homoclinic tangency. In this chapter we discuss the main ideas of the proof, leaving a complete presentation of it to Appendix V.

We recall that the stable set $W^s(\Lambda)$ of a basic set Λ is the union of the stable leaves through points of Λ. In the present context of surface diffeomorphisms, which we now assume to be at least C^2, this stable foliation is C^1. So the essential structure of $W^s(\Lambda)$ appears in its intersection with a curve ℓ transverse to the stable leaves—since the foliation is C^1, Hausdorff dimension and limit capacity of $W^s(\Lambda) \cap \ell$ are independent of ℓ (and they are equal to each other as we saw in the previous chapter). In particular we may take, for a saddle point $p \in \Lambda$, $\ell = W^u(p)$. We then define the *stable limit capacity* or *Hausdorff dimension* $d^s(\Lambda)$ as $d(W^s(\Lambda) \cap W^u(p))$; the *unstable limit capacity* is similarly defined.

Similar results for *heteroclinic cycles* as well as corresponding results in *higher dimensions* are stated in the last section. We also state a partial converse concerning nonhyperbolicity of maximal invariant sets when the above Hausdorff dimensions are *large* (sum bigger than 1).

As mentioned before, the result when the Hausdorff dimension is *small* (sum less than 1) becomes more of a *global nature* when the homoclinic tangency occurs as first bifurcation and the positive or negative limit set of the bifurcating diffeomorphism is hyperbolic: for a set of relatively large measure in the parameter space the corresponding diffeomorphisms have *their global limit set hyperbolic*.

The first result concerning relative measure of bifurcation sets (but not dealing with fractal dimensions) was obtained in [**NP**,1976], where homoclinic bifurcations from Morse–Smale diffeomorphisms was treated. The result we present here when applied to that case is somewhat stronger.

§1 Construction of bifurcating families of diffeomorphisms

We begin this section by indicating an interesting example of a persistently hyperbolic diffeomorphism on the 2-sphere S^2 with infinitely many periodic orbits: the diffeomorphism itself as well as every small perturbation of it has a hyperbolic limit set. It is in fact similar to the horseshoe example from Chapter 1 but it is constructed in $S^2 \cong \mathbb{R}^2 \cup \infty$ instead of in $\mathbb{R}^2$. Many other examples, and in fact a full discussion about constructing a homoclinic tangency as a "first" dynamic bifurcation, can be found in Appendix V.

We take in S^2 the diffeomorphic image of a square Q with two semicircular discs D_1 and D_2 attached as indicated in Figure 5.1.

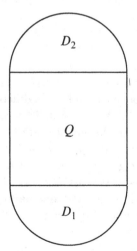

Figure 5.1

We let φ map $Q \cup D_1 \cup D_2$ inside itself as indicated in Figure 5.2, i.e. so that in Q we have the above horseshoe example and in D_1 we have one hyperbolic sink S_1, attracting all points in D_1.

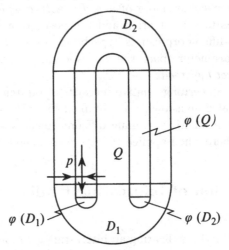

Figure 5.2

We extend φ to the complement of $Q \cup D_1 \cup D_2$ in S^2 in such a way that there is only one hyperbolic source S_0 and such that for each $x \in S^2 - (Q \cup D_1 \cup D_2)$, $\lim_{n \to \infty} \varphi^{-n}(x) = S_0$. It is easy to verify that in this case the positive limit set of φ consists of S_1, S_0 and the maximal invariant subset in Q. This last set can be analysed as in Chapter 2 and it is hyperbolic.

We let p denote the fixed saddle point in Q as indicated in Figure 5.2 and let Λ denote the maximal invariant subset of Q. The Cantor sets $W^s(p) \cap \Lambda$ and $W^u(p) \cap \Lambda$ are clearly dynamically defined (if φ is C^3 and if we take a correct identification of $W^s(p)$ and $W^u(p)$ with $\mathbb{R}$); these are the Cantor sets to which the results of Chapter 4 will be applied.

The first bifurcating family of diffeomorphisms that we construct in this section is based on the above example of a diffeomorphism φ of S^2 with a horseshoe, a source, and a sink. For simplicity we assume φ to be of class C^∞, although most results in this chapter are true for C^2 diffeomorphisms.

Let c be a curve from $r_s \in W^s(p)$ to $r_u \in W^u(p)$ as indicated in Figure 5.3. U denotes a small neighbourhood of c which is divided by the local components of $W^s(p) \cap U$ and $W^u(p) \cap U$ containing r_s and r_u in the regions U_I, U_{II}, and U_{III}. We shall obtain our one-parameter family by modifying the map φ in U, i.e. by composing φ with Ψ_μ, Ψ_μ a one-parameter family of diffeomorphisms which are, outside U, equal to the identity.

Before we describe Ψ_μ, we analyse the dynamic properties of orbits passing through U; we assume that this neighbourhood U of c is sufficiently small

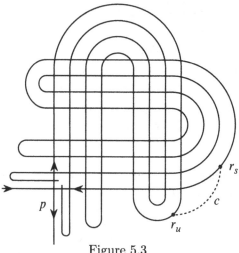

Figure 5.3

so that the following considerations are valid; see Figure 5.4. If $x \in U_I \cup U_{II}$ then $\varphi^n(x)$ tends, for $n \to +\infty$, to the sink S_1 and if $x \in U_{II} \cup U_{III}$ then $\varphi^n(x)$ tends, for $n \to -\infty$, to the source S_0. For $x \in U_{III}$, the positive iterates $\varphi^n(x)$, $n \to +\infty$, will stay near $W^u(p)$, but apart from that they may go to the sink or may stay near Λ (notice that $\Lambda = \overline{W^u(p) \cap W^s(p)}$; see the construction of φ above); in any case there are points $x \in U_{III}$ such that $\varphi^n(x) \in U_I$ for some positive n. Similarly for $x \in U_I$, the negative iterates $\varphi^{-n}(x)$, $n \to +\infty$, will stay near $W^s(p)$, but apart from that they may go to the source or may stay near Λ; in any case there are points $x \in U_I$ such that $\varphi^{-n}(x) \in U_{III}$ for some $n > 0$.

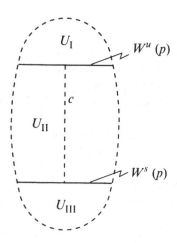

Figure 5.4

Now we come to the description of the one-parameter family Ψ_μ moving the points in U. We take Ψ_μ so that

- for $\mu \leq -1, \Psi_\mu$ is the identity,
- for $\mu > -1, \Psi_\mu$ pushes points down in U (in the direction of U_{III}) so that, for $\mu < 0$, U_I is still mapped inside $U_I \cup U_{II}$,
- for $\mu = 0$ there is a tangency of $\Psi_0(W^u(p))$ and $W^s(p)$, or more precisely of $\Psi_0(\overline{U}_I \cap \overline{U}_{II})$ and $\overline{U}_{II} \cap \overline{U}_{III}$; this tangency has quadratic order of contact and unfolds generically for $\mu > 0$ into two transversal intersections.

In Figure 5.5 we indicate the stable and unstable manifolds of p for the diffeomorphism $\Psi_\mu \circ \varphi$, $\mu = 0$.

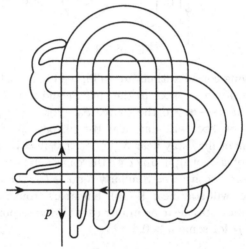

p

Figure 5.5

From the discussion of the dynamics of the points in U under iterations of φ, it follows that for $\mu < 0$, the positive limit set of $\varphi_\mu = \Psi_\mu \circ \varphi$ is the same as the positive limit set of φ. This implies that the bifurcation set of the one-parameter family $\varphi_\mu = \Psi_\mu \circ \varphi$ is contained in $\overline{R}_+ = \{\mu | \mu \geq 0\}$. In this case we say (as in Chapter 0) that the homoclinic tangency for $\mu = 0$ is a *first bifurcation*.

For $\mu = 0$, $L^+(\varphi_0) = L^+(\varphi)$, but now $L^+(\varphi_0)$ has a cycle. For $\mu > 0$, the positive limit set of φ_μ is strictly bigger than the positive limit set of φ. The orbits in $L^+(\varphi_\mu) - L^+(\varphi)$ are limits of orbits which go through U. From the above discussion of the dynamics of points in U under iteration of φ, it follows even that these φ_μ-orbits have to pass infinitely often through $\Psi_\mu(U_I) \cap U_{III}$.

We observe that the main properties of the one-parameter family φ_μ remain after a small perturbation of it (small in the C^∞ topology), the first bifurcation point remaining close to zero. In fact, we have constructed a whole open class of bifurcating arcs of diffeomorphisms and our results below are applicable to any element of this class.

Let Λ_μ denote the horseshoe of φ_μ, i.e. for $\mu \leq 0$, Λ_μ is the positive limit set of φ_μ excluding the source and the sink; for $\mu > 0$, Λ_μ is the continuation of Λ_0 which can be obtained, for small positive μ, by taking a sufficiently small neighbourhood V of Λ_0 and then taking Λ_μ the maximal invariant subset of V for φ_μ. For φ_μ as constructed, Λ_μ is independent of μ. But this does not persist under small perturbations of the one-parameter family. So, in general we must allow for a μ-dependence of Λ and the same can be said about the μ-dependence of the saddle point p.

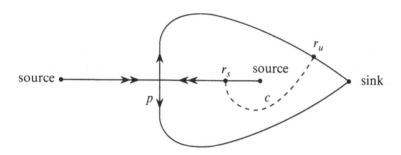

Figure 5.6

We now construct a second class of examples of bifurcating families for which our global result applies as well. Again on S^2, we consider a diffeomorphism φ whose limit set consists of two sources, one saddle and one sink as in Figure 5.6. The double arrow indicates that $\lambda\sigma < 1$, where $0 < \lambda < 1$ and $\sigma > 1$ are the eigenvalues of $(d\varphi)_p$. We say that λ is the dominating eigenvalue. Exactly as in the first example, we only modify φ in a small neighbourhood of the curve c joining the points r_u in $W^u(p)$ and r_s in $W^s(p)$ as in the figure. We define Ψ_μ as before and $\varphi_\mu = \Psi_\mu \circ \varphi$ so that we obtain a homoclinic orbit of tangency θ for $\varphi_0 = \Psi_0 \circ \varphi$. Using that $\lambda\sigma < 1$, one can check that $L^-(\varphi_0) = L(\varphi)$ and $L^+(\varphi_0) = L(\varphi) \cup \theta$. This fact was not present in the first example, since there we had $L^+(\varphi_0) = L^-(\varphi_0) = L(\varphi)$. We also point out that, in constructing this last example, we made explicit use of the eigenvalue condition $\lambda\sigma < 1$ vis-à-vis the geometry of the situation. If we repeat the construction but now with $\lambda\sigma > 1$, we get a family $\tilde{\varphi}_\mu$ for which there are cascades of bifurcations as in Section 3 of Chapter 3, for small values of μ both positive and negative! Still, even in this case, our results concerning hyperbolicity or the lack of it when unfolding a homoclinic tangency apply. The proofs are essentially the same. However, in this general case, in the absence of global data (say a filtration), we focus our attention on the maximal invariant set in the union of a neighbourhood of the homoclinic orbit of tangency and a neighbourhood of the basic set associated to it. In this discussion we are assuming $\lambda\sigma \neq 1$. Indeed, it is an interesting question how to adapt our results to the conservative (area preserving) case.

We finish this section with some comments on the stable and unstable foliations and the curve of tangency between their leaves. As in Chapter 2, or Appendix 1, we construct the stable and unstable foliations $\mathcal{F}_\mu^s$ and $\mathcal{F}_\mu^u$ on a neighbourhood of Λ_μ. These foliations are $C^{1+\varepsilon}$ and depend differentiably on μ, which means that the *tangent directions* to the leaves depend $C^{1+\varepsilon}$ on both the space and parameter variables; see Appendix 1. We can extend the domains of definition of $\mathcal{F}_\mu^s$ and $\mathcal{F}_\mu^u$ by applying φ_μ^{-1}, respectively φ_μ. In this way we can obtain that both $\mathcal{F}_\mu^s$ and $\mathcal{F}_\mu^u$ are defined in a neighbourhood of the point of first tangency in U. In particular we want both $\mathcal{F}_\mu^s$ and $\mathcal{F}_\mu^u$ for μ positive and small to be defined on the disc bounded by the two components of $W^s(p_\mu) \cap U$ and $W^u(p_\mu) \cap U$ which made the first tangency. From now on we denote these components, which are also leaves of $\mathcal{F}_\mu^s$ and $\mathcal{F}_\mu^u$, by $\tilde{W}^s(p_\mu)$ and $\tilde{W}^u(p_\mu)$. (See Figure 5.7).

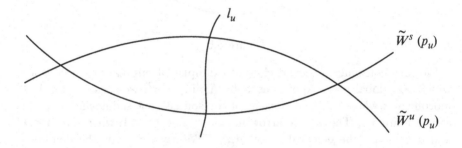

Figure 5.7

The curve ℓ_μ is defined as the set of points where the leaves of $\mathcal{F}_\mu^s$ and $\mathcal{F}_\mu^u$ are tangent. Since the contact between $\tilde{W}^u(p_0)$ and $\tilde{W}^s(p_0)$ is parabolic and since the tangent directions to the leaves of the foliations $\mathcal{F}_\mu^u$ and $\mathcal{F}_\mu^s$ are C^1, it follows that ℓ_μ is a differentiable curve depending differentiably on μ.

In both $\mathcal{F}_\mu^s$ and $\mathcal{F}_\mu^u$ there is a collection of exceptional leaves—the stable and unstable manifolds of points of Λ_μ. The collection of these leaves is denoted by $\mathcal{F}^s(\Lambda_\mu), \mathcal{F}^u(\Lambda_\mu)$: they are just the leaves of $\mathcal{F}_\mu^s, \mathcal{F}_\mu^u$ passing through points of Λ_μ.

In the following sections we mainly deal with the geometry of the configuration formed by $\mathcal{F}_\mu^s, \mathcal{F}_\mu^u, \mathcal{F}^s(\Lambda_\mu), \mathcal{F}^u(\Lambda_\mu), \ell_\mu$, when restricted to the disc bounded by $\tilde{W}^s(p_\mu)$ and $\tilde{W}^u(p_\mu)$.

§2   Homoclinic tangencies with bifurcation set of small relative measure – statement of the results

The results which we discuss in this section can be applied to the classes of one-parameter families constructed in the last section and we obtain the following consequence.

THEOREM 1. *Let φ_μ be the first one-parameter family as constructed in the last section. Let d^u and d^s be the (local) limit capacity of $W^s(p_0) \cap \Lambda_0$ and $W^u(p_0) \cap \Lambda_0$. If $d^u + d^s < 1$, then*

$$\lim_{\mu_0 \to 0} \frac{m(B(\varphi_\mu) \cap [0, \mu_0])}{\mu_0} = 0,$$

where $B(\varphi_\mu)$ is the bifurcation set of the one-parameter family φ_μ: $B(\varphi_\mu) = \{\mu \in \mathbb{R} | \varphi_\mu$ is not Ω-stable\}, and where $m(\ \ )$ denotes the Lebesgue measure.

In the next section we shall describe the structure of the proof of this theorem which is based on [**PT**,1987]. Here, in this section, we discuss much more general assumptions under which the conclusion of the theorem still holds.

First, for simplicity we consider one-parameter families of C^∞ diffeomorphisms, but in general the results are valid for C^2 diffeomorphisms. Also, the ambient manifold M is a compact surface without boundary, but extensions of the results to higher dimensions are possible; see Section 4 of the present chapter.

Second, in the presence of small Hausdorff dimensions (limit capacities) we shall distinguish two possibilities: we get small relative measure of the bifurcation set *either* when the diffeomorphisms in the family are considered on all of M (global bifurcations) *or* when they are restricted to some neighbourhood of the orbit of homoclinic tangency and the associated basic set. In this last case, we consider bifurcations of the maximal invariant set of the restriction of the map.

So, we begin by considering one-parameter families φ_μ having one (and only one) homoclinic orbit of tangency, say θ, for $\mu = 0$, which is quadratic and unfolds generically, and besides $\mu = 0$ is a first bifurcation, i.e. for $\mu < 0$ the limit set $L(\varphi_\mu)$ is persistently hyperbolic. Let $L^+(\varphi_\mu)$ be the positive limit set of φ_μ. We assume that $\lim_{\mu \nearrow 0} L^+(\varphi_\mu) = L_0^+$ exists and is hyperbolic for φ_0 exhibiting no cycles except for θ. This means that the homoclinic tangency should not take place in the limit L_0^+ of positive limit sets. Moreover, we *assume that* $L^+(\varphi_0) = L_0^+$; *this implies for the nonwandering set the relation* $\Omega(\varphi_0) = L_0^+ \cup \{\theta\}$.

These restrictions, *or similar ones for the negative limit set*, are imposed by the methods of proof used in this *global setting* as we shall discuss in the next section. Probably only some of these conditions are really necessary for our main conclusion: *small relative measure of the global bifurcation set near zero.*

In this context it is worth commenting on the example of a homoclinic bifurcation as indicated in Figure 5.8 (see Section 4, Chapter 1): a horseshoe map composed with a *translation downwards*. It is not known in general whether the first tangency of stable and unstable manifold of p_μ is a first bifurcation value of the parameter. If it is a first bifurcation, it is happening with a tangency *inside* the positive (and also negative) limit set and thus not like the example we have constructed in the previous section. The reason why the orbit of tangency is inside the limit set is the following: it is accumulated by transverse homoclinic orbits and, so, by Chapter 2, also by periodic orbits. We have no information on the relative measure of the global bifurcation set in this case.

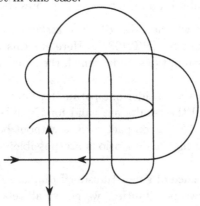

Figure 5.8

For a discussion of these assumptions and the restrictions which they impose on the global dynamics of φ_0 and the topology of M, see [**PT**,1987]. Also, in some cases depending on the global dynamics, we have to require that $|\det(d\varphi_0)_{p_0}| \neq 1$, where p_0 denotes the saddle point related to the homoclinic tangency. (For the first family constructed in the previous section one did not have to impose this last condition on the determinant of $d\varphi_0$). The limit capacities of the restrictions of the stable and unstable foliations to the basic set (or simply stable and unstable limit capacities) are defined as the (local) limit capacities of $W^u(p_0) \cap L_0^+$ and $W^s(p_0) \cap L_0^+$. As before, L_0^+ is the limit of $L^+(\varphi_\mu)$ for μ approaching 0 from below and p_0 is the saddle point of φ_0 associated with the homoclinic tangency. Notice, as seen in Chapter 4, that the limit capacities of the above sets are the same as their Hausdorff dimensions.

Then under these more general assumptions we have the same conclusion.

THEOREM 2. *In the above situation, if the sum of stable and unstable limit capacities is smaller than 1, then*

$$\lim_{\mu_0 \to 0} \frac{m(B(\varphi_\mu) \cap [0, \mu_0])}{\mu_0} = 0.$$

We now present essentially the same result but in more general terms. The main novelty is that we restrict the dynamics to some neighbourhood of the orbit of homoclinic tangency and its associated basic set. But, on the other hand, the assumptions are more general: we only assume the homoclinic tangency to be quadratic and to unfold generically. In particular, an important case is that of one-parameter families of *dissipative* diffeomorphisms (i.e. area contracting) going through generic homoclinic bifurcations; see the Hénon-like maps at the end of Chapter 3 and also Chapter 6. Notice that here we are not mentioning any hypothesis about the homoclinic tangency (say at $\mu = 0$) being a first bifurcation for the family φ_μ.

The neighbourhood U_μ to which we now restrict the dynamics of φ_μ is of the form $U_\mu = U_1 \cup U_\mu^*$, where

- U_1 is a small fixed neighbourhood (independent of μ) of the associated basic set say Λ_0, so that Λ_0 is the maximal invariant set for φ_0 in U_1,

- $U_\mu^* = \bigcup_{|i| \leq N} \varphi_\mu^i(V_\mu)$, V_μ is a neighbourhood of a point in the orbit of tangency whose points are at a distance at most $K\mu$ from both the local stable and unstable manifolds of p_μ, for some constant $K > 0$.

We want N above to be the smallest integer so that $U_1 \cup U_\mu^*$ is a neighbourhood of the homoclinic orbit of tangency. Let us denote by Γ_μ the maximal invariant set of $\varphi_\mu | U_\mu$. We then say that μ_1 is a bifurcating point or parameter value for the family $\{\varphi_\mu | U_\mu\}$ if there exists μ_2 arbitrarily close to μ_1 such that φ_{μ_1} restricted to a neighbourhood of Γ_{μ_1} is not conjugate to φ_{μ_2} restricted to a neighbourhood of Γ_{μ_2}. Denote by $B(\{\varphi_\mu | U_\mu\})$ or simply $B(\varphi_\mu | U_\mu)$ this set of bifurcating points of the family $\{\varphi_\mu | U_\mu\}$. We can now state the following general result.

THEOREM 3. *For a family φ_μ as above, if the sum of the limit capacities (Hausdorff dimensions) of Λ_0 is smaller than 1, then*

$$\lim_{\mu_0 \to 0} \frac{m(B(\varphi_\mu | U_\mu) \cap [-\mu_0, \mu_0])}{\mu_0} = 0$$

This holds for any $K > 0$ in the above definition of U_μ.

Note that the choice of a larger constant K leads to a slower rate of convergence of the limit.

§3   Homoclinic tangencies with bifurcation set of small relative measure – idea of proof

In this section we want to outline the proof of the first theorem in the previous section. The proof of the more general and even global theorem is the same except for a careful analysis of the consequences of the requirement that φ_0 is a *first bifurcation* in relation with the eigenvalues of $d\varphi_0$ at p_0; we conclude this section with a discussion of this last point. A detailed exposition and proofs of these results are in [**PT**,1987], which is reproduced here in Appendix 5. However, the present outline indicates some of the main ideas and may be useful as a preparation for reading the complete proof. Finally, the last theorem is actually a *localization* of the previous one to a neighbourhood of the homoclinic orbit of tangency and of the associated basic set.

We use the notation introduced in Section 2. It is clear that for any μ, such that there is an orbit of tangency of $\mathcal{F}^s(\Lambda_\mu)$ and $\mathcal{F}^u(\Lambda_\mu)$, φ_μ is a bifurcating diffeomorphism (either such an orbit belongs to the positive limit set $L^+(\varphi_\mu)$ which is then nonhyperbolic, or $L^+(\varphi_\mu)$ has a cycle). From the considerations in Section 2 it follows that any such orbit of tangency has a point in the disc bounded by $\tilde{W}^s(p_\mu)$ and $\tilde{W}^u(p_\mu)$ and hence on the line ℓ_μ of tangencies (tangencies not on ℓ_μ may appear when extending the domains of $\mathcal{F}^s_\mu$ and $\mathcal{F}^u_\mu$, and hence of $\mathcal{F}^s(\Lambda_\mu)$ and $\mathcal{F}^u(\Lambda_\mu)$, through iterations of φ_μ^{-1}, respectively φ_μ). So, in order to prove that, for some μ, φ_μ is hyperbolic we need at least $\mathcal{F}^s(\Lambda_\mu) \cap \ell_\mu$ and $\mathcal{F}^u(\Lambda_\mu) \cap \ell_\mu$ to be disjoint, in fact we need their distance to be bigger than some lower bound depending on μ. Notice that we use here $\mathcal{F}^s(\Lambda_\mu)$ and $\mathcal{F}^u(\Lambda_\mu)$ also to denote the union of their leaves; this is usually called the stable, respectively unstable, set of Λ_μ and denoted by $W^s(\Lambda_\mu)$, $W^u(\Lambda_\mu)$.

PROPOSITION 1. *For each $c > 0$, there is a $\mu_1(c) > 0$, such that for every $\mu \in (0, \mu_1(c))$ such that the distance between $\mathcal{F}^s(\Lambda_\mu) \cap \ell_\mu$ and $\mathcal{F}^u(\Lambda_\mu) \cap \ell_\mu$ is at least $c \cdot \mu$, φ_μ is hyperbolic.*

The proof of this proposition is based on the fact that the *new* orbits of $L^+(\varphi_\mu)$, i.e. the orbits of $L^+(\varphi_\mu) - L^+(\varphi)$ are all contained in a small neighbourhood of $\mathcal{F}^s(\Lambda_\mu) \cap \mathcal{F}^u(\Lambda_\mu)$. More precisely, one has for a suitable choice of $\mu_1(c)$ (in Proposition 1) that for any $\mu \in (0, \mu_1(c))$ and any $z \in L^+(\varphi_\mu) - L^+(\varphi)$, there is a point $\tilde{z}$ in the orbit of z such that

- $\tilde{z}$ is enclosed by $\tilde{W}^s(p_\mu)$ and $\tilde{W}^u(p_\mu)$,

- $\tilde{z}$ is within distance $1/4 \cdot c \cdot \mu$ from both $\mathcal{F}^s(\Lambda_\mu)$ and $\mathcal{F}^u(\Lambda_\mu)$.

The proof of the above fact is based on an argument involving filtrations (see Smale [**S**,1970], Conley [**C**,1978]) which implies that the new orbits of $L^+(\varphi_\mu)$, $\mu > 0$, have to pass through a small neighbourhood U_μ of Λ_μ: small in the sense that $\lim_{\mu \searrow 0} U_\mu = \Lambda_0$. As observed in Chapter 0, the existence of filtrations yields this kind of global control on the limit set or nonwandering set. From this first fact it follows that for $\mu \in (0, \mu_1(c))$ if the distance between $\mathcal{F}^s(\Lambda_\mu) \cap \ell_\mu$ and $\mathcal{F}^u(\Lambda_\mu) \cap \ell_\mu$ is at least $c \cdot \mu$, then the new points of $L^+(\varphi_\mu)$ which are near ℓ_μ, are in the region where the leaves of $\mathcal{F}^s_\mu$ and $\mathcal{F}^u_\mu$ are transverse. One can even derive a lower bound for the angles between leaves of these foliations in this region which is of order $(c\mu)^{1/2}$. This transversality, together with the fact that these new orbits of $L^+(\varphi_\mu)$ must stay a long time (increasing to infinity for $\mu \searrow 0$) near the hyperbolic set Λ_μ before they return to a neighbourhood of ℓ_μ, enables us to prove the hyperbolicity of $L^+(\varphi_\mu)$ using the construction of cone fields as in Section 3 of Chapter 2. This concludes the outline of the proof of Proposition 1.

Next we have to indicate why for $d^s + d^u < 1$ and small $c > 0$, the subset of $(0, \mu_1(c))$ of those μ for which the distance between $\mathcal{F}^s(\Lambda_\mu) \cap \ell_\mu$ and $\mathcal{F}^u(\Lambda_\mu) \cap \ell_\mu$ is at least $c \cdot \mu$ is of big relative measure.

To deduce this we need some preparation; see Figure 5.9. First, we identify ℓ_μ with $\mathbb{R}$ (or at least with a part of $\mathbb{R}$) such that $0 \in \mathbb{R}$ corresponds with $\tilde{W}^s(p_\mu) \cap \ell_\mu$ and $\mu \in \mathbb{R}$ corresponds with $\tilde{W}^u(p_\mu) \cap \ell_\mu$. We denote the sets in $\mathbb{R}$, corresponding to $\mathcal{F}^s(\Lambda_\mu) \cap \ell_\mu$ and $\mathcal{F}^u(\Lambda_\mu) \cap \ell_\mu$, by A_μ and B_μ, respectively.

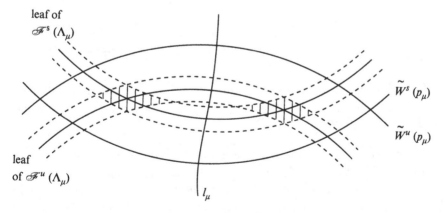

leaf of $\mathcal{F}^s(\Lambda_\mu)$

leaf of $\mathcal{F}^u(\Lambda_\mu)$

$\tilde{W}^s(p_\mu)$

$\tilde{W}^u(p_\mu)$

ℓ_μ

shaded area : contains points of possible "new" orbits of $L^+(\varphi_\mu)$ near ℓ_μ

Figure 5.9

Without loss of generality we may assume that distances in ℓ_μ correspond to differences in $\mathbb{R}$. The foliations $\mathcal{F}^s_\mu$ and $\mathcal{F}^u_\mu$ are C^1; see Appendix 1 for their differentiability, also with respect to μ. Thus the sets A_μ and B_μ are diffeomorphic images of $W^u(p_\mu) \cap \Lambda_\mu$ and $W^s(p_\mu) \cap \Lambda_\mu$ and, hence, have limit capacities d^s and d^u, respectively. See Figure 5.10 for the relative positions of A_μ and B_μ.

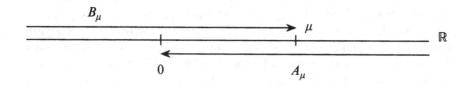

Figure 5.10

As a second step, we have the following result relating the sets A_μ, B_μ with A_0 and $B_0 + \mu$. The proof of this result is technical and hard to explain in a brief way, so we just refer the reader to Appendix 5.

PROPOSITION 2. *For each $c > 0$ there is a $\mu_2(c) > 0$ such that for each $\mu \in (0, \mu_2(c))$, $A_\mu \cap [0, \mu]$ and $B_\mu \cap [0, \mu]$ are contained in a $(c \cdot \mu)$-neighbourhood of A_0 and $B_0 + \mu$, respectively.*

This proposition allows us to obtain the necessary estimates on the distances between A_μ and B_μ from estimates on the distances between A_0 and $B_0 + \mu$ as given below.

PROPOSITION 3. *Let ρ be some constant. Consider the set $B^{(\bar\mu, c)} = \{\mu \in (0, \bar\mu) |$ distance between A_0 and $B_0 + \mu$ is less than $c \cdot \mu\}$. If $d^u + d^s = d(A_0) + d(B_0) < 1$, then for each $\varepsilon > 0$ there is $c > 0$ such that for any $0 < \bar\mu < \rho$,*

$$\frac{m(B^{(\bar\mu, c)})}{\bar\mu} < \varepsilon,$$

where m denotes Lebesgue measure.

PROOF: We choose $d_1 > d^u = d(A_0)$ and $d_2 > d^s = d(B_0)$ so that $d_1+d_2 < 1$. Since A_0 and B_0 are diffeomorphic images of *scaled* sets we have, for c sufficiently small, that $A_0 \cap [0, \overline{\mu}]$ and $B_0 \cap [-\overline{\mu}, 0]$ can be covered with c^{-d_1} respectively c^{-d_2} intervals of length $c \cdot \overline{\mu}$ (for this to be true, uniformly in $\overline{\mu}$, we needed to bound $\overline{\mu}$ by ρ). So a $(c \cdot \overline{\mu})$-neighbourhood of $A_0 \cap [0, \overline{\mu}]$ can be covered with not more than $3 \cdot c^{-d_1}$ intervals of length $c \cdot \overline{\mu}$. This means that

$$\tilde{B}^{(\overline{\mu},c)} = \{\mu \in (0, \overline{\mu})| \text{ distance between } A_0 \text{ and } B_0 + \mu \text{ is less than } c \cdot \overline{\mu}\},$$

being the difference (see Chapter 4) of $B_0 \cap [-\overline{\mu}, 0]$ and a $(c \cdot \overline{\mu})$-neighbourhood of $A_0 \cap [0, \overline{\mu}]$, can be covered with not more than $3 \cdot c^{-d_1-d_2}$ intervals of length $c \cdot \overline{\mu}$. So the measure of $\tilde{B}^{(\overline{\mu},c)}$ is at most $3 \cdot c^{1-d_1-d_2} \cdot \overline{\mu}$. Since $d_1 + d_2 < 1$, $c^{1-d_1-d_2} \to 0$ for $c \to 0$, and the proposition follows from the fact that $\tilde{B}^{(\overline{\mu},c)} \supset B^{(\overline{\mu},c)}$. $\qquad\square$

Finally, we show how the first theorem of the previous section follows from the above three propositions. For fixed $\varepsilon > 0$, we show that for μ_0 sufficiently small,

$$\frac{m(B(\varphi_\mu) \cap [0, \mu_0])}{\mu_0} < \varepsilon. \tag{1}$$

We only consider values of μ (or μ_0) which are smaller than the constant ρ in Proposition 3. We choose c_1, using Proposition 3, such that

$$\frac{m(B^{(\overline{\mu},c_1)})}{\overline{\mu}} < \varepsilon$$

for all $\overline{\mu} \in (0, \rho)$. Then, using Propositions 1 and 2, we find $\overline{\mu}_0$ such that

- for every $\mu \in (0, \overline{\mu}_0)$ such that the distance between $\mathcal{F}^s(\Lambda_\mu) \cap \ell_\mu$ and $\mathcal{F}^u(\Lambda_\mu) \cap \ell_\mu$ is at least $\frac{1}{2} \cdot c_1 \cdot \mu$, φ_μ is hyperbolic,
- for every $\mu \in (0, \overline{\mu}_0)$, $A_\mu \cap [0, \mu]$ and $B_\mu \cap [0, \mu]$ are contained in a $(\frac{1}{4} \cdot c_1 \cdot \mu)$-neighbourhood of A_0, respectively $B_0 + \mu$.

The inequality (1) holds for all $\mu_0 < \overline{\mu}_0$. In fact, for such μ_0, $(B(\varphi_\mu) \cap [0, \mu_0]) \subset B^{(\mu_0, c_1)}$ and the Lebesgue measure of the last set is at most $\varepsilon \cdot \mu_0$. This ends the outline of the proof of theorem 1 of Section 2. $\qquad\square$

We conclude this section by presenting further explanation of the condition $|\det(d\varphi_0)_{p_0}| \neq 1$ mentioned in Section 2, in relation with the hypothesis of φ_0 being the first bifurcation. The above outline of proof, in which this condition on the determinant of $d\varphi_0$ at p_0 was not used, referred to the first one-parameter family constructed in Section 1. In that one-parameter family, all the *new* orbits of $L^+(\varphi_\mu)$ have a point in the disc bounded by $\tilde{W}^s(p_\mu)$ and $\tilde{W}^u(p_\mu)$. This gives a first limitation of the region which we have to take into account in the rest of the arguments. The situation is, however, sometimes more complicated, e.g. consider the example of a two-dimensional diffeomorphism with homoclinic tangency indicated in Figure 5.11 and like the second one-parameter family constructed in Section 1.

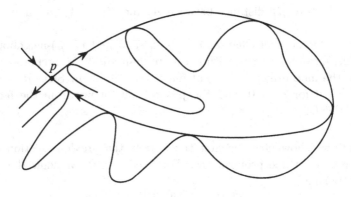

Figure 5.11

Let φ_μ be a generic one-parameter family, with φ_0 being a diffeomorphism as above, unfolding a homoclinic tangency between $W^u(p)$ and $W^s(p)$ which turns into two transerve intersections for $\mu > 0$. We then have several possibilities.

– If $|\det(d\varphi(p))| > 1$ then $\mu = 0$ is *not* a first bifurcation value (and hence this one-parameter family does not satisfy the assumptions formulated in Section 2): the horseshoes as constructed in Section 4, Chapter 3, appear for $\mu < 0$. Still, the last (semi-global) theorem in Section 2 applies.

– If $|\det(d\varphi(p))| < 1$ then $\mu = 0$ is a first bifurcation value, but the new points of $L^+(\varphi_\mu)$ which are near the tangency are *not* enclosed by the analogue of $\tilde{W}^s(p_\mu)$ and $\tilde{W}^u(p_\mu)$ as in Section 3. However, it is still possible in this case to show that all new points of $L^+(\varphi_\mu)$, for $\mu > 0$ small, have to pass near the orbit of tangency. And they have even to pass through a neighbourhood of a point in the orbit of tangency, consisting of the points at distance less than $K \cdot \mu$ from both the local arcs in $W^s(p)$ and $W^u(p)$ involved in the tangency, for some constant K. So, also here we get a first limitation of the region which can contain new points of $L^+(\varphi_\mu)$.

- If $|\det(d\varphi(p))| = 1$, which in particular occurs for conservative diffeo-
morphisms, our global methods do not work but the last (semi-global)
theorem in Section 2 does apply. Of course, in our setting this is an
ungeneric case between the above two.

This discussion indicates how the global dynamics at the first bifurcation
leads to the condition on the determinant of $d\varphi_0$ at p_0, if we want to obtain
results about the *global limit set or chain recurrent set* of φ_μ, μ small.

§4   Heteroclinic cycles and further results on measure of bifurcation sets

We discuss here extensions of the theorems on unfolding of homoclinic tangencies and measure of bifurcation sets in Sections 2 and 3 of this chapter.
   We formulate the first result in terms of the positive limit set L^+; a similar
statement can be made in terms of the negative limit set L^-. We consider
a family of diffeomorphisms φ_μ on M^2 and assume that φ_μ is hyperbolic for
$\mu \leq 0$; i.e. $L^+(\varphi_\mu)$ is a hyperbolic set for all $\mu \leq 0$. As we saw in Chapter
0, we can write

$$L^+(\varphi_0) = \Lambda_1 \cup \cdots \cup \Lambda_s$$

where each Λ_i is a basic set. Again we assume that we have a cycle Γ
among the basic sets in $L^+(\varphi_0)$, but now we assume that the cycle involves
more than one basic set, say $j > 1$ of them, $\Lambda_{i_1}, \ldots, \Lambda_{i_j}$. Recall that the
notion of cycle was introduced in Chapter 0. *Following Chapter 4 and the
introduction to Chapter 5, we define the stable and unstable limit capacities
of the cycle* $\Gamma = \{\Lambda_{i_1}, \ldots, \Lambda_{i_j}\}$ *as*

$$d^s(\Gamma) = \max_{1 \leq k \leq j} d^s(\Lambda_{i_k}),$$
$$d^u(\Gamma) = \max_{1 \leq k \leq j} d^u(\Lambda_{i_k}).$$

Similar to the families of surface diffeomorphisms in the previous sections
of the present chapter, we consider φ_μ such that

(i) the positive limit set of φ_μ is hyperbolic for $\mu \leq 0$ and has the no-cycle
property for $\mu < 0$,

(ii) φ_0 has a cycle Γ with a unique orbit of tangency between stable and
unstable manifolds, which is quadratic and unfolds generically.

As before, let $B(\varphi_\mu) = \{\mu \in \mathbb{R} \mid \varphi_\mu \text{ is not } \Omega\text{-stable}\}$ and let $m(\quad)$ denote
the Lebesgue measure. We then have the following result of a global nature.

THEOREM 1 [**PT**,1985]. *For a family* $\{\varphi_\mu\}$ *as above, if* $d^s(\Gamma) + d^u(\Gamma) < 1$,
then

$$\lim_{\mu_0 \to 0} \frac{m(B(\varphi_\mu) \cap [0, \mu_0])}{\mu_0} = 0.$$

This result is essentially in [**PT**,1985], but with a different proof from the
one we indicated in Section 3 of the present chapter. Actually, in that paper
it is proved that the lim inf of the above expression is zero, instead of the full
limit. This is an extension of a result in [**NP**,1976], where all basic sets are
just periodic orbits. The reason one can now prove the stronger statement
is the following fact, shown in [**T**,1988], which enables us to apply the proof
in [**PT**,1987] for homoclinic bifurcations outlined in the previous section.
Suppose that the tangency in the cycle Γ occurs between the unstable manifold of a fixed (periodic) point p_{i_1} in Λ_{i_1} and the stable manifold of a fixed
(periodic) point p_{i_j} in Λ_{i_j}. Then we can prove that

$$d\left(\left(\bigcup_{k=1}^{j} W^u(\Lambda_{i_k})\right) \cap W^s(p_{i_1})\right) = d^u(\Gamma)$$

and

$$d\left(\left(\bigcup_{k=1}^{j} W^s(\Lambda_{i_k})\right) \cap W^u(p_{i_j})\right) = d^s(\Gamma),$$

where as before, d stands for limit capacity. Indeed, first recall from Chapter
0 that the unstable (stable) set $W^u(\Lambda_{i_k})$ $(W^s(\Lambda_{i_k}))$ of a basic set as Λ_{i_k} is
the union of the unstable (stable) manifolds of the points in Λ_{i_k}. Now
we observe that the Cantor sets that appear in the left-hand side of the
expressions above, if some of the basic sets are nontrivial, are not of the type
we have called *dynamically defined*. However, one can still prove that their
limit capacity is equal to their Hausdorff dimension. From this the equalities
above follow since they are true in general for Hausdorff dimensions.

Figure 5.12 indicates examples of cycles as above, Figure 5.12 (a) involving
two fixed saddles as basic sets and Figure 5.12 (b) involving a horseshoe and
a saddle point.

Keeping the same notation, we still assume conditions (i) and (ii)
above for the family φ_μ, but now we are concerned with *global* (structural) stability of φ_μ for $\mu > 0$ and small. Then, to begin with, we
must assume the *transversality condition* for all stable and unstable manifolds of φ_0 except along the (unique) orbit of tangency occurring say in
$W^u(\Lambda_{i_1}) \cap W^s(\Lambda_{i_j})$. Besides, one has also to consider the unstable limit
capacities of the basic sets Λ_k such that $W^u(\Lambda_k) \cap W^s(\Lambda_{i_1}) \neq \phi$. Each such
Λ_k is said to be *positively involved with the tangency* since $W^u(\Lambda_k)$ contains

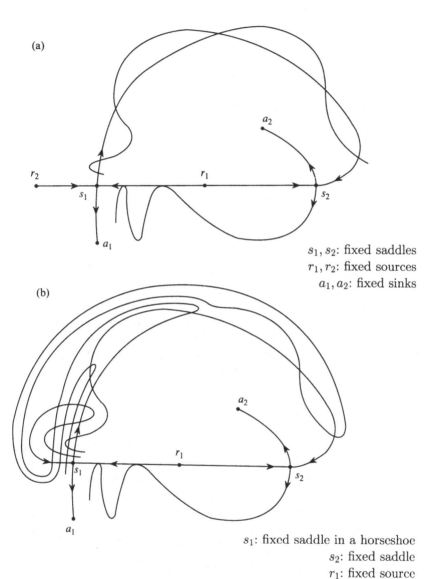

(a)

a_2

r_2 s_1 r_1 s_2

a_1

s_1, s_2: fixed saddles
r_1, r_2: fixed sources
a_1, a_2: fixed sinks

(b)

a_2

r_1

s_1 s_2

a_1

s_1: fixed saddle in a horseshoe
s_2: fixed saddle
r_1: fixed source
a_1, a_2: fixed sinks

Figure 5.12

the orbit of tangency in its closure. Similarly, Λ_k is *negatively involved with the tangency* if $W^s(\Lambda_k)$ intersects $W^u(\Lambda_{i_j})$. We define the *unstable limit capacity of the tangency* d^u_t as the maximum of the unstable limit capacities of the basic sets positively involved with the tangency. Similarly, we can define the *stable limit capacity of the tangency* d_t^s. For a periodic orbit which is a basic set we define these limit capacities as zero even when they are sources or sinks. On the other hand for an attractor which is not periodic the stable limit capacity is 1; similarly for a repeller which is not periodic.

So, for a family φ_μ as above we have the following.

THEOREM 2. *If* $d_t^s + d_t^u < 1$, *then*

$$\lim_{\mu_0 \to 0} \frac{m(B_g(\varphi_\mu) \cap [0, \mu_0])}{\mu_0} = 0$$

where $B_g(\varphi_\mu) = \{\mu \in \mathbb{R} \mid \varphi_\mu$ *is not globally stable*$\}$.

As in Section 2 of this chapter, we can state a general result concerning families φ_μ that go through a cycle say Γ at $\mu = 0$. Again, we do not have to assume complete hyperbolicity of $L^+(\varphi_0)$ or $L^-(\varphi_0)$ but just that the cycle Γ occurs among basic sets. Also, we do not assume any condition about $\mu = 0$ being a first bifurcation or Ω-bifurcation for φ_μ but just that the tangency is quadratic and unfolds generically. However, we then have to restrict ourselves to a neighbourhood U_μ of the orbit of tangency together with a small fixed neighbourhood of Γ; see Sections 2 and 3. The bifurcation set $B(\varphi_\mu)$ is now replaced by the set $B(\varphi_\mu \mid U_\mu)$ of values of μ for which φ_μ restricted to a neighbourhood of the maximal invariant set of $\varphi_\mu \mid U_\mu$ is not stable; see Section 2 of this chapter. Then, if $d_\Gamma^s + d_\Gamma^u < 1$ we have

$$\lim_{\mu_0 \to 0} \frac{m(B(\varphi_\mu \mid U_\mu) \cap [-\mu_0, \mu_0])}{\mu_0} = 0.$$

We now consider a generalization of the previous results on homoclinic tangencies to *higher dimensions*. Let φ_μ be a family of C^∞ diffeomorphisms on an n-dimensional compact boundaryless manifold M, $n \geq 2$. Let Λ_0 be a basic set for φ_0 with a fixed (periodic) point p and an associated homoclinic tangency in $(W^s(p) \cap W^u(p)) - \Lambda_0$. As before, we assume the tangency to be *quadratic* and to *unfold generically*. Furthermore, now we want $W^u(p)$ and $W^{cs}(p)$ as well as $W^s(p)$ and $W^{cu}(p)$ to be transverse along the orbit of tangency, where $W^{cs}(p)$ and $W^{cu}(p)$ stands for the centre stable and centre unstable manifolds; see Appendix 1. For an overall view of these generic conditions see [**NPT**,1983]. We also assume that the weakest contracting and expanding eigenvalues have *multiplicity 1* and, although this may not be necessary, that they are *real*. Moreover, we suppose that $\Lambda_0 \cap W^{ss}(p) = \phi$ and $\Lambda_0 \cap W^{uu}(p) = \phi$, where $W^{ss}(p)$, and $W^{uu}(p)$ are the strong stable and unstable manifolds; see Appendix 1. We consider in $W^s(p)$ the strong stable foliation $\mathcal{F}^{ss}(p)$, whose leaves have codimension 1, which contains $W^{ss}(p)$ as a leaf and is of class C^∞ [**HPS**,1977]. Similarly, we have in $W^u(p)$ the strong unstable foliation $\mathcal{F}^{uu}(p)$. These foliations define C^∞ projections π^s and π^u on one-dimensional manifolds ℓ_s and ℓ_u (the spaces of leaves). We then *define* the *unstable* and *stable limit capacities* $d^u(\Lambda_0, p)$ and $d^s(\Lambda_0, p)$ as the limit capacity of $\pi^s(\Lambda_0 \cap W^s(p))$ and of $\pi^u(\Lambda_0 \cap W^u(p))$, respectively.

For a family φ_μ as above, if $d^u(\Lambda_0, p) + d^s(\Lambda_0, p) < 1$, then according to [**T**,1991a] (see also [**T**,1988] and [**T**,1991b]) the previous results can be adapted to yield

$$\lim_{\mu \to 0} \frac{m(B(\varphi_\mu \mid U_\mu) \cap [-\mu_0, \mu_0])}{\mu_0} = 0.$$

As before, U_μ is a neighbourhood of the orbit of tangency together with a small fixed neighbourhood of Λ_0, and $B(\varphi_\mu \mid U_\mu)$ indicates the set of bifurcation values of φ_μ restricted to a neighbourhood of the maximal invariant set of $\varphi_\mu \mid U_\mu$.

A similar result may be formulated in higher dimensions for heteroclinic cycles which are equidimensional, i.e. all the stable manifolds of the basic sets have the same dimension. In the case where all basic sets are just periodic orbits, the result was obtained in [**NP**,1976] but with lim inf *instead of full limit in the above expression*.

As in Section 2 or earlier in this section, we may also impose the usual conditions on φ_0 being a first bifurcating diffeomorphism and $L^+(\varphi_0)$ (or $L^-(\varphi_0)$) being hyperbolic. Then, if $d^u(\Lambda_0, p) + d^s(\Lambda_0, p) < 1$, we obtain a set of relatively large measure near $\mu = 0$ corresponding to *globally stable or Ω-stable diffeomorphisms*.

Finally, we return to homoclinic tangencies (or cycles) for families of *surface diffeomorphisms* φ_μ. Let Λ_0 be a basic set for φ_0, $p \in \Lambda_0$ a fixed (periodic) point with an associated homoclinic tangency. Now we assume $HD^s(\Lambda_0) + HD^u(\Lambda_0) > 1$ (or equivalently $d^s(\Lambda_0) + d^u(\Lambda_0) > 1$). Then, from [**PY**,1991], we have the following converse to the theorems above.

THEOREM 3. *For almost all families φ_μ as above we have*

$$\limsup_{\mu_0 \to 0} \frac{m(B(\varphi_\mu \mid U_\mu) \cap [-\mu_0, \mu_0])}{\mu_0} > 0.$$

This and the previous results convey a fundamental difference in the measure of the bifurcation set depending on how large or how small is the Hausdorff dimension of the basic set associated to the homoclinic bifurcation: small (< 1) dimension implies "few" dynamic bifurcations and big (> 1) dimension is likely to imply "much" dynamic bifurcation.

CHAPTER 6

INFINITELY MANY SINKS AND
HOMOCLINIC TANGENCIES

In this chapter we deal with a surprising phenomenon discovered by New-
house [**N**,1970]: there is an open subset $U \subset \text{Diff}^2(M)$, M a 2-manifold, in
which the set of diffeomorphisms exhibiting a homoclinic tangency is dense.
He showed later [**N**,1974] that, under some additional hypotheses, such a
set U has a residual subset $R \subset U$ such that each $\varphi \in R$ has infinitely many
hyperbolic periodic attractors (sinks). This disproved a conjecture of Thom
that generically a diffeomorphism or a flow only has finitely many attractors.
These results are treated in the next section of this chapter which follows
essentially the original papers [**N**,1970], [**N**,1974].

Later it was realized that this phenomenon was actually very common or
abundant in the presence of an initial homoclinic tangency: whenever $\varphi \in$
Diff $^2(M)$, $\dim(M) = 2$, has a saddle point p with a *homoclinic tangency*
of $W^u(p)$ and $W^s(p)$, then *arbitrarily close to* φ there is an open set $U \subset$
Diff $^2(M)$ such that the diffeomorphisms which have a homoclinic tangency
are dense in U, [**N**,1979]. If, moreover, $|\det(d\varphi)_p| < 1$, one can take this
open subset U so that it has a *residual subset* $R \subset U$ with the property that
each $\varphi' \in R$ has *infinitely many hyperbolic periodic attractors*. The main
purpose of this chapter is to prove this last statement. This result can be
adapted to one-parameter families of surface diffeomorphisms going through
a homoclinic tangency bifurcation [**R**,1983]. We also present a new proof of
this fact in Appendix 4.

The proof presented here of this important result differs in several ways
from the original paper [**N**,1979]. It is a pleasure to acknowledge Newhouse's
fundamental role in this new proof of his result: in an oral communication
with the present authors he provided arguments and constructions (see Section 2 on the tent map and the logistic map and, to some extent, Section
3) which made it possible to bypass what we thought could be a gap in
his original proof. When writing the present proof, these last ideas of Newhouse became central: they enabled us to make the whole construction more
geometric.

Recently, the results in this chapter have been generalized to higher dimensions [**PV**,1991]; see Chapter 7.

§1 Persistent tangencies

In this section we consider the phenomenon of open subsets $U \subset$ Diff $^2(M)$, M a 2-manifold, such that for each $\varphi \in U$, there are tangencies or almost tangencies of the stable and unstable manifolds of (hyperbolic) basic sets of φ. To be more explicit, we say that $U \subset$ Diff $^2(M)$ has *persistent tangencies* if there are continuous maps Λ_1 and Λ_2 which assign to each φ in U basic sets $\Lambda_1(\varphi)$ and $\Lambda_2(\varphi)$ of saddle type, and there is a tangency between $W^u(x_1)$ and $W^s(x_2)$ for some $x_1 \in \Lambda_1(\varphi)$ and $x_2 \in \Lambda_2(\varphi)$ or such a tangency can be obtained by an arbitrarily C^2-small perturbation of φ. The results will then be applied to get *persistent homoclinic tangencies*: this is just the case where $\Lambda_1 = \Lambda_2$.

In the construction of persistent tangencies we shall have to deal with the notion of thickness for basic sets. For Cantor sets, this notion is defined in Chapter 4.

Let $\varphi \in \text{Diff}^2(M)$ be a two–dimensional diffeomorphism, let Λ be a basic set of saddle type for φ and let $p \in \Lambda$ be a fixed (or periodic) point. One can parametrize $W^s(p)$ and $W^u(p)$ in such a way that $\varphi \mid W^s(p)$ and $\varphi \mid W^u(p)$ become linear. In $W^s(p)$ and $W^u(p)$ we shall measure distances with respect to such a linearizing parametrization. We define the *unstable* and the *stable thickness* of (Λ, p) as $\tau^u(\Lambda, p) = \tau(W^s(p) \cap \Lambda)$ and $\tau^s(\Lambda, p) = \tau(W^u(p) \cap \Lambda)$. Observe that since $W^s(p) \cap \Lambda$ is φ-invariant and since $\varphi \mid W^s(p)$ is linear, there are arbitrarily small compact neighbourhoods K of p in $W^s(p) \cap \Lambda$ such that $\tau(K) = \tau(W^s(p) \cap \Lambda) = \tau^u(\Lambda, p)$. The same applies to $\tau^s(\Lambda, p)$. It can be shown (but we shall not use this fact here) that $\tau^s(\Lambda, p)$ is independent of p: if p' is another periodic point in Λ then $\tau^s(\Lambda, p') = \tau^s(\Lambda, p)$. A similar statement holds for the unstable thickness. Finally, we again recall from Section 3, Chapter 4, that the stable and unstable thicknesses depend continuously on φ with respect to the C^2 topology. This means that if $U \subset \text{Diff}^2(M)$ is open and if, for $\varphi \in U$, $\Lambda(\varphi)$ is a basic set of saddle type with fixed (or periodic) point $p(\varphi)$, both depending continuously on $\varphi \in U$, then both $\tau^s(\Lambda(\varphi), p(\varphi))$ and $\tau^u(\Lambda(\varphi), p(\varphi))$ are continuous functions on U.

Now we give the first type of persistent tangencies.

PROPOSITION 1. *Let $\varphi \in \text{Diff}^2(M)$ be a C^2 diffeomorphism on a surface M with basic sets Λ_1 and Λ_2, both of saddle type, and let $p_1 \in \Lambda_1$ and $p_2 \in \Lambda_2$ be fixed or periodic points. We assume that $\tau^u(\Lambda_1, p_1) \cdot \tau^s(\Lambda_2, p_2) > 1$ and that there is an orbit of tangency of $W^u(p_1)$ and $W^s(p_2)$. Then φ is in the closure of some $U \subset \text{Diff}^2(M)$, where U has persistent tangencies involving the continuations $\Lambda_1(\tilde{\varphi})$ and $\Lambda_2(\tilde{\varphi})$ of Λ_1 and Λ_2 for $\tilde{\varphi} \in U$.*

PROOF: For a sufficiently small neighbourhood V of φ in $\text{Diff}^2(M)$ we define the maps $\Lambda_1, \Lambda_2, p_1$ and p_2 which assign to $\tilde{\varphi} \in V$ the basic sets $\Lambda_1(\tilde{\varphi})$ and

$\Lambda_2(\tilde{\varphi})$ and the fixed or periodic points $p_1(\tilde{\varphi})$ and $p_2(\tilde{\varphi})$, depending continuously on $\tilde{\varphi}$ and coinciding for $\tilde{\varphi} = \varphi$ with $\Lambda_1, \Lambda_2, p_1$, and p_2 as introduced above. Also, we define an unstable foliation $\mathcal{F}_1^u(\tilde{\varphi})$ for $\Lambda_1(\tilde{\varphi})$ and a stable foliation $\mathcal{F}_2^s(\tilde{\varphi})$ for $\Lambda_2(\tilde{\varphi})$, depending continuously on $\tilde{\varphi} \in V$, as in Chapter 2, Section 3 and Chapter 4. We assume now that the tangency of $W^u(p_1)$ and $W^s(p_2)$ is quadratic. Otherwise we can obtain this with an arbitrarily small perturbation. Near a point of the orbit of tangency of $W^u(p_1)$ and $W^s(p_2)$ we have (for $\tilde{\varphi}$ sufficiently close to φ) a C^1 line $\ell(\tilde{\varphi})$ of tangencies of $\mathcal{F}_1^u(\tilde{\varphi})$ and $\mathcal{F}_2^s(\tilde{\varphi})$. Also there are $\tilde{\varphi}$-dependent projections

$$\pi_1(\tilde{\varphi}): W^s(p_1(\tilde{\varphi})) \to \ell(\tilde{\varphi})$$

and

$$\pi_2(\tilde{\varphi}): W^u(p_2(\tilde{\varphi})) \to \ell(\tilde{\varphi})$$

which project along leaves of $\mathcal{F}_1^u(\tilde{\varphi})$ and $\mathcal{F}_2^s(\tilde{\varphi})$, respectively. Here, $\ell(\tilde{\varphi})$ and $\pi_1(\tilde{\varphi})$, $\pi_2(\tilde{\varphi})$ are C^1 and depend continuously on $\tilde{\varphi} \in V$.

Let $K_1(\tilde{\varphi})$, $K_2(\tilde{\varphi})$ be small compact one-sided neighbourhoods of $p_1(\tilde{\varphi})$ and $p_2(\tilde{\varphi})$ in $W^s(p_1(\tilde{\varphi})) \cap \Lambda_1(\tilde{\varphi})$ and $W^u(p_2(\tilde{\varphi})) \cap \Lambda_2(\tilde{\varphi})$ depending continuously on $\tilde{\varphi} \in V$, and such that $\tau(K_1(\tilde{\varphi})) = \tau^u(\Lambda_1(\tilde{\varphi}), p_1(\tilde{\varphi}))$ and $\tau(K_2(\tilde{\varphi})) = \tau^s(\Lambda_2(\tilde{\varphi}), p_2(\tilde{\varphi}))$. We define $L_i(\tilde{\varphi}) = (\pi_i(\tilde{\varphi}))(K_i(\tilde{\varphi}))$ for $i = 1, 2$. Since thickness is not in general preserved by diffeomorphisms, we cannot expect $\tau(L_i(\tilde{\varphi}))$ to be equal to $\tau(K_i(\tilde{\varphi}))$. On the other hand, thickness is better preserved by a (one-dimensional) diffeomorphism when the ratio of the maximal and minimal values of its derivative is closer to 1. So, by taking K_1 and K_2 sufficiently small, and also taking the neighbourhood V of φ small, we can achieve that for all $\tilde{\varphi} \in V$, $\tau(L_1(\tilde{\varphi})) \cdot \tau(L_2(\tilde{\varphi})) > 1$. Finally, we take $U \subset \text{Diff}^2(M)$ as $U = \{\tilde{\varphi} \in V \mid L_1(\tilde{\varphi}) \text{ and } L_2(\tilde{\varphi}) \text{ are linked in } \ell(\tilde{\varphi})\}$ (linked Cantor sets are defined in Section 2, Chapter 4). Clearly, U is open. Since $L_1(\varphi)$ and $L_2(\varphi)$ have a boundary point in common, φ is in the closure of U. For $\tilde{\varphi} \in U$, $L_1(\tilde{\varphi})$ and $L_2(\tilde{\varphi})$ intersect each other and hence there is a tangency of $W^u(\Lambda_1(\tilde{\varphi}))$ and $W^s(\Lambda_2(\tilde{\varphi}))$. This completes the proof of the proposition. $\qquad \square$

The above proposition is often applied in a somewhat different form. The two corollaries below are variations of the type which we shall use later in this chapter.

COROLLARY 1. *Assume in proposition 1 that $\varphi \in \text{Diff}^2(M)$ has basic sets Λ_1 and Λ_2 of saddle type with fixed (or periodic) points $p_1 \in \Lambda_1$ and $p_2 \in \Lambda_2$ and that p_0 is another saddle point of φ such that $W^s(p_0)$ and $W^u(p_1)$ have a transverse intersection while $W^u(p_0)$ and $W^s(p_2)$ have an orbit of tangency. If $\tau^u(\Lambda_1, p_1) \cdot \tau^s(\Lambda_2, p_2) > 1$, then we have the same conclusion.*

PROOF: This follows from the fact that $W^u(p_1)$ accumulates on $W^u(p_0)$ and, hence, arbitrarily close to a diffeomorphism φ as in the corollary, there is a diffeomorphism with an orbit of tangency between $W^u(p_1)$ and $W^s(p_2)$. To such diffeomorphisms we can apply our proposition. □

COROLLARY 2. *Let $\varphi, \Lambda_1, \Lambda_2, p_1$ and p_2 be as in the above proposition, i.e. $W^u(p_1)$ and $W^s(p_2)$ have an orbit of tangency and $\tau^u(\Lambda_1, p_1) \cdot \tau^s(\Lambda_2, p_2) > 1$. Suppose also that φ has a basic set Λ_3 of saddle type with fixed (or periodic) point $p_3 \in \Lambda_3$ such that $W^u(p_3)$ and $W^s(p_1)$ have a transverse intersection. Then the conclusion of the proposition still holds if we replace the continuation $\Lambda_1(\tilde{\varphi})$ of Λ_1 by the continuation $\Lambda_3(\tilde{\varphi})$ of Λ_3.*

PROOF: As a consequence of the transverse intersection of $W^u(p_3)$ and $W^s(p_1)$, each leaf of $W^u(\Lambda_1)$ is accumulated by leaves of $W^u(\Lambda_3)$. So, whenever some $\tilde{\varphi}$ near φ has a tangency between $W^u(\Lambda_1(\tilde{\varphi}))$ and $W^s(\Lambda_2(\tilde{\varphi}))$, we can obtain a tangency between $W^u(\Lambda_3(\tilde{\varphi}'))$ and $W^s(\Lambda_2(\tilde{\varphi}'))$ for a diffeomorphism $\tilde{\varphi}'$ which is arbitrarily close to $\tilde{\varphi}$. □

Now we consider *persistent homoclinic tangencies* which are just persistent tangencies between stable and unstable leaves of (the continuation of) the *same* basic set. Due to the above proposition, we get persistent homoclinic tangencies arbitrarily near a diffeomorphism $\varphi \in \mathrm{Diff}^2(M)$ if $\tau^s(\Lambda, p) \cdot \tau^u(\Lambda, p) > 1$ and if $W^u(p)$ and $W^s(p)$ have a homoclinic tangency, where Λ is a basic set for φ and $p \in \Lambda$ is a fixed (periodic) point.

As we have seen in Chapter 3, we can perturb a two–dimensional diffeomorphism φ with a homoclinic tangency associated to a saddle point p with $|\det(d\varphi)_p| < 1$ so as to create a hyperbolic periodic attractor (sink). Moreover, the orbit of the sink that we can create passes arbitrarily near the original tangency. Combining this idea with the construction of persistent homoclinic tangencies we obtain the following proposition.

PROPOSITION 2. *Let $U \subset \mathrm{Diff}^2(M)$ be an open set of two–dimensional diffeomorphisms with persistent homoclinic tangencies, associated with a basic set $\Lambda(\varphi)$. Let $p(\varphi) \in \Lambda(\varphi)$ be a periodic point, say of period k and let $|\det(d\varphi^k)_{p(\varphi)}| < 1$. Then there is a residual subset $R \subset U$ such that each $\varphi \in R$ has infinitely many hyperbolic periodic attractors (sinks). If $|\det(d\varphi^k)_{p(\varphi)}| > 1$, one gets infinitely many periodic repellers (sources).*

PROOF: It is enough to show that whenever $\varphi \in U$ has n hyperbolic periodic attractors, $n \geq 0$, then there is arbitrarily near φ a diffeomorphism $\varphi' \in U$ with $n + 1$ hyperbolic periodic attractors. So we take such $\varphi \in U$ with n hyperbolic periodic attractors, which we denote by $\mathcal{O}_1(\varphi), \ldots, \mathcal{O}_n(\varphi)$. Since a hyperbolic periodic orbit is persistent under small perturbations of the diffeomorphism, we can consider $\mathcal{O}_i(\cdot)$, $i = 1, \ldots, n$, as continuous maps defined on a small neighbourhood V of φ in U, assigning to each $\tilde{\varphi} \in V$ the n hyperbolic periodic attractors $\mathcal{O}_1(\tilde{\varphi}), \ldots, \mathcal{O}_n(\tilde{\varphi})$. By assumption, we have for φ, or some φ' close to φ, a tangency between $W^u(\Lambda)$ and $W^s(\Lambda)$. Since $W^u(p)$ is dense in $W^u(\Lambda)$ and $W^s(p)$ is dense in $W^s(\Lambda)$, we may even approximate φ by a diffeomorphism φ' such that $W^u(p(\tilde{\varphi}'))$ and $W^s(p(\tilde{\varphi}'))$ have a tangency. Let q be a point of such an orbit of tangency. Since $q \notin \mathcal{O}_1(\varphi') \cup \cdots \cup \mathcal{O}_n(\varphi')$, there are neighbourhoods W of q and V' of φ' such that for $\varphi'' \in V'$, $(\mathcal{O}_1(\varphi'') \cup \cdots \cup \mathcal{O}_n(\varphi'')) \cap W = \phi$. By the results of Chapter 3, there is, arbitrarily close to φ', a diffeomorphism φ'' which has a hyperbolic periodic attractor containing a point of W. Hence, such φ'' has at least $n + 1$ hyperbolic periodic attractors. This proves the proposition.

□

REMARK 1: As consequence of the arguments in the above proof, we also get the following result. Let φ_μ be a one-parameter family of diffeomorphisms as in Chapter 3, having for $\mu = 0$ a homoclinic tangency (parabolic and unfolding generically) associated to the saddle point p_0. Suppose that $|\det(d\varphi_0)_{p0}| < 1$ and that p_μ is part of a nontrivial basic set Λ_μ such that $W^u(\Lambda_\mu)$ and $W^s(\Lambda_\mu)$ have persistent tangencies for μ near 0. Then, for generic μ close to 0, φ_μ has infinitely many sinks.

§2   The tent map and the logistic map

In this section we deal with one-dimensional maps. The results can be applied to diffeomorphisms near a homoclinic tangency due to the result in Section 4, Chapter 3, according to which such diffeomorphisms *contain* arbitrarily close approximations of the one-dimensional logistic map.

First we define the *tent map* $T : [0, 1] \to [0, 1]$ by

$$T(x) = \begin{cases} 2x & \text{if} \quad 0 \le x \le \frac{1}{2}, \\ 2 - 2x & \text{if} \quad \frac{1}{2} \le x \le 1. \end{cases}$$

We prove that T has arbitrarily thick invariant Cantor sets. In fact we shall construct for each $m \ge 3$ a Cantor set $K_m \subset [0, 1]$ such that

- the map T, restricted to a neighbourhood of K_m, is differentiable, i.e. $\frac{1}{2} \notin K_m$, and hence $dT \mid K_m$ is uniformly expanding,
- the thickness of $K_m \to \infty$ as $m \to \infty$, and in fact $\tau(K_m) = 2^{m-1} - 3$.

Construction of K_m. We fix $m \ge 3$ and construct K_m. Let $q \in (0, 1)$ be the point of period m for T whose orbit $q = q_0, q_1 = T(q), \ldots, q_m = T^m(q) =$

q satisfies

$$q_2 < q_3 < \cdots < q_{m-1} < \frac{1}{2} < q_m = q_0 < q_1.$$

The case $m = 4$ is illustrated in Figure 6.1. For each $m \geq 3$ there is exactly one such point q. Let $q_m^* \in (0, \frac{1}{2})$ be defined by $T(q_m^*) = T(q)$.

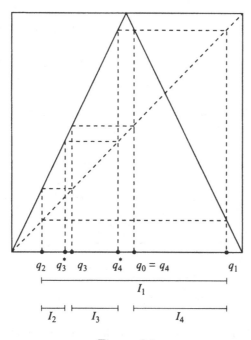

$$q_2 \quad \overset{*}{q_3} \quad q_3 \qquad \overset{*}{q_4} \quad q_0 = q_4 \qquad\qquad q_1$$

Figure 6.1

For $i = 3, \ldots, m - 1$ we define q_i^* by $q_i^* = T^{-1}(q_{i+1}^*) \cap (0, \frac{1}{2})$. With these points we define the intervals $I_1 = [q_2, q_1]$, $I_2 = [q_2, q_3^*]$, ..., $I_{m-1} = [q_{m-1}, q_m^*]$, and $I_m = [q_m, q_1]$. So $I_i \subset I_1$ for $i = 2, \ldots, m$ and T maps I_i to I_{i+1} for $i = 2, \ldots, m - 1$ and maps I_m to I_1. Define $K_m^{(1)} = I_2 \cup \cdots \cup I_m$ and $K_m^{(n)} = \bigcap_{i=0}^{n-1} T^{-i}(K_m^{(1)})$. The Cantor set K_m is defined as $K_m = \lim_{n \to \infty} K_m^{(n)}$. So K_m is a dynamically defined Cantor set in the sense of Chapter 4.

It follows immediately from the construction that K_m is invariant under T, it does not contain the point $\frac{1}{2}$ and $\{T^{-i}(q_0)\}$ is dense in K_m. Now we come to the calculation of the thickness of K_m. Let δ be the distance between q_m^* and q_m (we shall later express δ in terms of m). Then $q_1 = 1 - \delta$ and $q_2 = 2\delta$, while $q_m = \frac{1}{2} + \frac{1}{2}\delta$ and $q_m^* = \frac{1}{2} - \frac{1}{2}\delta$. So the length of I_1 is $1 - 3 \cdot \delta$. The biggest gap in K_m is the interval (q_m^*, q_m) of length δ. To the right of this gap there is I_m of length $\frac{1}{2} - \frac{3}{2} \cdot \delta$ and to the left of this gap is the interval $[q_2, q_m^*]$

of length $\frac{1}{2} - \frac{5}{2} \cdot \delta$. So, the minimum of the thicknesses of K_m at the boundary points of its biggest gap (q_m^*, q_m) is $\frac{1}{2\delta} - \frac{5}{2}$. For any other gap U of K_m there is a closed interval I containing U and an integer n such that $T^n \mid I$ is a linear map onto I_1, such that $T^n(U) = (q_m^*, q_m)$. This implies that for each gap we get the same local thickness. So $\tau(K_m) = \frac{1}{2\delta} - \frac{5}{2}$. Finally, we express δ in terms of m. As we saw, $q_2 = 2 \cdot \delta$. From the construction it follows that $T^{m-2}(q_2) = q_m = \frac{1}{2} + \frac{1}{2} \cdot \delta$, and that $T^i(q_2) < \frac{1}{2}$ for $i = 1, \ldots, m-3$. So, $2^{m-2} \cdot 2 \cdot \delta = \frac{1}{2} + \frac{1}{2} \cdot \delta$ or $\delta = \frac{1}{2^m - 1}$. From this it follows that $\tau(K_m) = 2^{m-1} - 3$. This concludes our construction.

Next we define the *logistic map* $L: [0, 1] \to [0, 1]$ as $L(x) = 4x(1 - x)$. Also for this map we want to construct thick invariant Cantor sets on which the derivative of L, or of some iterate L^n of L, is uniformly expanding. These Cantor sets will be obtained through a conjugacy between L and T from the Cantor sets K_m which we just constructed.

Geometrically, the easiest way to see this conjugacy between the tent and the logistic maps is to consider both as two-to-one projections of the square map of the circle S^1. The square map is defined as $e^{i\varphi} \mapsto (e^{i\varphi})^2 = e^{i \cdot 2 \cdot \varphi}$ or $\varphi \mapsto 2\varphi \mod 2\pi$. This is indicated in the following diagram:

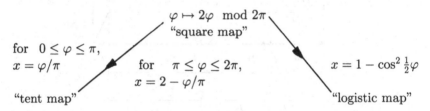

$$\varphi \mapsto 2\varphi \mod 2\pi$$
"square map"

for $0 \le \varphi \le \pi$,
$x = \varphi/\pi$

for $\pi \le \varphi \le 2\pi$,
$x = 2 - \varphi/\pi$

$x = 1 - \cos^2 \frac{1}{2}\varphi$

"tent map"   "logistic map"

Both the projections, from the square map to the tent map and the logistic map, are easily seen to be equivariant, i.e. semi-conjugacies. This means that $x \mapsto 1 - \cos^2 \frac{1}{2}\pi x$ defines a conjugacy from the tent map to the logistic map; we denote this map by Ψ.

The thick invariant Cantor sets for the logistic map that we want are $\tilde{K}_m = \Psi(K_m)$, where K_m are the invariant Cantor sets for the tent map we have just constructed. Thus, $\tilde{K}_m$ is clearly invariant. Since Ψ, restricted to a neighbourhood of K_m, is a diffeomorphism, $\tilde{K}_m$ is hyperbolic. That is, there are constants $A > 0$ and $\lambda > 1$, depending on m, such that $(L^n)'(x) \ge A \cdot \lambda^n$ for $x \in \tilde{K}_m$ and $n \ge 1$. Let us see why they have large thickness.

We recall that the thickness is defined as the infimum of ratios of length of intervals, as explained in Section 2, Chapter 4. If we transform a Cantor set K by a diffeomorphism and if $\alpha \ge 1$ is the ratio between the maximum and the minimum of the norm of the derivative of the diffeomorphism, then the thickness of the image Cantor set is at least $\alpha^{-1} \cdot \tau(K)$. In our case the *diffeomorphism* is $\Psi(x) = 1 - \cos^2 \frac{1}{2}\pi x$ and $\Psi'(x) = \frac{\pi}{2} \cdot \sin \pi x$, which is zero for $x = 0$ and $x = 1$. Although K_m does not contain 0 or 1, the sets

K_m, $m \geq 3$, are not uniformly bounded away from 0 and 1. So, we cannot yet conclude that the thickness of $\tilde{K}_m$ goes to infinity. For this, we need the next two observations.

(1) For small fixed $\varepsilon > 0$, $\Psi \mid [\varepsilon, 1 - \varepsilon]$ is a diffeomorphism, and so here we can apply the above reasoning with a constant α (depending on ε).

(2) Near 0, we have the following fact. If $U = (u_1, u_2)$ is a gap of K_m (near zero), then the worst deformation occurs if we compare the length ratios of (u_1, u_2), $(0, u_1)$ and $(u_2, 2u_1)$ (since $(2u_1, 2u_2)$ is another gap of K_m) with the corresponding length ratios after applying Ψ. Since, for x near 0, Ψ is essentially quadratic, these length ratios do not differ by more than a factor, say 4. The same applies for gaps near 1.

Combining the above two points we find that $\tau(\tilde{K}_m) \geq (\overline{\alpha})^{-1} \cdot \tau(K_m)$ with $\overline{\alpha} = \max(\alpha, 4)$. Thus, $\tau(\tilde{K}_m)$ becomes arbitrarily big as $m \to \infty$.

REMARK 1: We point out that (independent of the above argument) the local thickness is preserved under diffeomorphisms and, for dynamically defined Cantor sets like K_m and $\tilde{K}_m$, it is the same in all points. So the local thickness of $\tilde{K}_m$ is at least $2^{m-1} - 3$, and it is this fact which we shall use later.

CONCLUSION 1: For each $m \geq 3$, there is a periodic point Q_m for the logistic map (with period m) such that

$$\tilde{K}_m = C\ell \bigcup_{i \geq 0} L^{-i}(\{Q_m, L(Q_m), \ldots, L^{m-1}(Q_m)\})$$

is a hyperbolic invariant Cantor set for L. The thicknesses satisfy the condition $\lim_{m \to \infty} \tau(\tilde{K}_m) = \infty$.

§3 Hénon-like diffeomorphisms

In this section we consider two–dimensional diffeomorphisms which are close to the endomorphisms

$$(x, y) \mapsto (y, y^2 + \mu)$$

obtained in Section 4, Chapter 3, and discuss some consequences of the constructions in the previous section. We shall consider a continuous two–parameter family φ_{μ_1, μ_2} of C^2 maps on $\mathbb{R}^2$ such that

$$\varphi_{\mu_1, 0}(x, y) = (y, y^2 + \mu_1)$$

and such that for $\mu_2 > 0$, φ_{μ_1, μ_2} is a diffeomorphism. Up to some reparametrizations one may think of the family $(x, y) \mapsto (1 - ax^2 +$

y, bx). This is known as the Hénon family: based on numerical experiments, he conjectured the existence of a strange attractor for such a family [**H**,1976]; see Chapter 7. The endomorphisms $\varphi_{\mu_1,0}$ are equivalent to the one-dimensional maps $y \mapsto y^2 + \mu_1$. If $\mu_1 = -2$, this last map is equivalent to the logistic map: consider the change of variables $x = -\frac{1}{4}y + \frac{1}{2}$. For the logistic map we constructed hyperbolic invariant sets $\tilde{K}_m$; here we investigate what remains of them for $\mu = (\mu_1, \mu_2)$ near $(-2, 0)$ with $\mu_2 > 0$.

In Figure 6.2 we indicate $\tilde{K}_3$ as invariant subset of $\varphi_{-2,0}$ in the curve $\ell = Im(\varphi_{-2,0})$, or at least the first stage of its construction: $\tilde{K}_3^{(1)} = I_2 \cup I_3$.

Note that the boundary points of I_2 and I_3 consist of the orbit of Q_3, as indicated in the figure, and one point of the inverse image of that orbit. In general, we have for each $m \geq 3$ a periodic point Q_m in ℓ such that the inverse images of its orbit are dense in $\tilde{K}_m$. So we use the same notation as in the last section, but now it refers to $\varphi_{-2,0} \mid \ell$ instead of L.

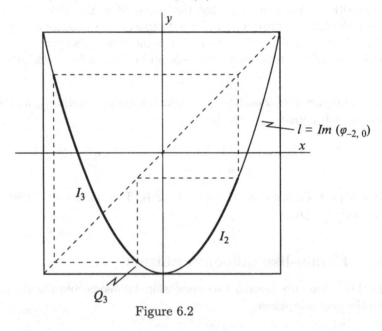

Figure 6.2

PROPOSITION 1. *For each $m \geq 3$ there is a neighbourhood W of $(-2, 0)$ in the (μ_1, μ_2)-half plane $\{(\mu_1, \mu_2) \mid \mu_2 \geq 0\}$ and continuous maps P, Q_m, and Λ_m which assign to each $\mu \in W$ the fixed point $P(\mu)$, the periodic point $Q_m(\mu)$ and the invariant set $\Lambda_m(\mu)$ for φ_μ, in such a way that for $\mu = (-2, 0)$ we have*

   − $P(-2, 0) = (2, 2)$,

   − *$Q_m(-2, 0)$ is the periodic point Q_m mentioned above,*

$- \Lambda_m(-2,0) = \tilde{K}_m,$

and such that for $\mu = (\mu_1, \mu_2)$ with $\mu_2 > 0$, $\Lambda_m(\mu)$ is a basic set containing $Q_m(\mu)$. Finally, for each m we have that the stable thickness of $(\Lambda_m(\mu), Q_m(\mu))$ is continuous at $\mu = (\mu_1, 0)$, i.e. $\tau^s(\Lambda_m(\mu), Q_m(\mu))$ converges to $\tau(\Lambda_m(\mu_1, 0))$ as $\mu \to (\mu_1, 0)$. In particular, $\tau^s(\Lambda_m(\mu), Q_m(\mu))$ can be made arbitrarily large by taking m large and μ near $(-2, 0)$.

PROOF: The existence of the functions P and Q_m follows from the implicit function theorem. We shall concentrate on Λ_m. In order to simplify the discussion we take the case $m = 3$, but the arguments apply in general. First we take small neighbourhoods U_2 and U_3 of I_2 and I_3 as indicated in Figure 6.3 (for $m > 3$ we must replace I_3 by I_m and I_2 by the minimal interval in ℓ which contains $I_2, \ldots, I_{m-1}$). In these neighbourhoods we have *horizontal* foliations whose leaves are inverse images of points by $\varphi_{-2,0}$. We assume that each component of a horizontal leaf in U_i has exactly one point of intersection with ℓ.

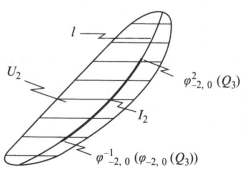

Figure 6.3

We define $\Lambda_3(\mu)$ to be the maximal invariant subset of φ_μ in $U_2 \cup U_3$. In order to complete the proof of the proposition, we need to analyse the dynamics of φ_μ restricted to $U_2 \cup U_3$ for μ near $(-2, 0)$.

First we prove that $\Lambda_3(\mu)$ is hyperbolic; see [**S**,1981] for similar results. For this we use cone fields as introduced in Section 3, Chapter 2. We describe how the unstable cone field is constructed, the stable cone field being complementary. In each point of $U_2 \cup U_3$ we take the unstable cone so that it does not contain the horizontal direction; in points of $\ell \cap (U_2 \cup U_3)$ we take the unstable cones so that they contain the tangent direction of ℓ in their interior. Since $\tilde{K}_3$ is an expanding hyperbolic set for $\varphi_{-2,0} \mid \ell$, we have, for $U_2 \cup U_3$ sufficiently small, constants $N \geq 1$ and $C > 1$ such that for each x, v with $x \in U_2 \cup U_3$ and $\varphi^i_{-2,0}(x) \in U_2 \cup U_3$ for $i = 1, \ldots, N$ and v in the unstable cone at x, $\left| d\varphi^N_{-2,0}(v) \right| \geq C \cdot |v|$. This implies that also for φ_μ, μ near $(-2, 0)$, the vectors in the unstable cones are eventually growing exponentially if they do not leave $U_2 \cup U_3$. The corresponding fact for the stable cones is evident. So this implies that $\Lambda_3(\mu)$ is hyperbolic. The fact that it is a basic set can be shown by constructing a Markov partition. This can be done as in Appendix 2, using the stable and unstable manifolds of the orbit of $Q_3(\mu)$: this Markov partition consists of two boxes which are near I_2 and I_3 respectively (in the case that $m > 3$ one has $m - 1$ boxes which are *continuations* of $I_2, \ldots, I_m$). This Markov partition is clearly mixing.

The continuity of the stable thickness is proved along the same lines as for diffeomorphisms; see Chapter 4. Still we give a short description of the proof.

First we need to show that the unstable leaves of $W^u(\Lambda_3(\mu))$ in $U_2 \cup U_3$ are C^2 curves which are C^2-close to ℓ. The union of these leaves in $U_2 \cup U_3$ is $\bigcap_{i \geq 0} \varphi^i_\mu(U_2 \cup U_3)$. For μ close to $(-2, 0)$ this is C^0-close to ℓ. For μ near $(-2, 0)$ the unstable cones can be chosen arbitrarily narrow. This means, since we know that the individual unstable leaves are C^1 and even C^2, that the unstable leaves are also C^1 near ℓ. Next we define $L(\mathbb{R}^2) = \{(x, L) \mid x \in \mathbb{R}^2, L$ a one-dimensional linear subspace of $T_x(\mathbb{R}^2)\}$. The map in $L(\mathbb{R}^2)$, induced by the derivative of φ_μ, is denoted by $D\varphi_\mu$. The curve $L(\ell) = \{(x, T_x(\ell) \mid x \in \ell\}$ is the $D\varphi_{-2,0}$ image of a neighbourhood of $L(\ell)$. Applying the above arguments to $D\varphi_\mu$ (which is C^1), we get a hyperbolic set for $D\varphi_\mu$ with unstable leaves C^1 close to $L(\ell)$. But for each unstable leaf F^u for φ_μ in $\mathbb{R}^2$ there is a corresponding unstable leaf $L(F^u) = \{(x, T_x(F^u)) \mid x \in F^u\}$ for $D\varphi_\mu$. This implies that the unstable leaves for φ_μ in $\mathbb{R}^2$ are C^2 near ℓ.

From the construction in the previous section, it is clear that there is for each $n \geq 1$ a compact part $W^s_n(\mu)$ in the stable manifold of the orbit of $Q_3(\mu)$ such that $W^s_n(\mu)$ intersects ℓ transversally and such that these intersections converge, for $\mu \to (-2, 0)$, to the boundary points of the intervals of $\tilde{K}^{(n)}_3 = \bigcap_{i=0}^{n-1} \varphi^{-i}_{-2,0}(I_2 \cup I_3)$. Since both $W^s_n(\mu)$ and the unstable foliation depend continuously on μ, the thickness of $W^u(Q_3(\mu)) \cap \Lambda_3(\mu)$, as far as *big* gaps are concerned, is continuous.

For the *small* gaps, we use the bounded distortion property for φ_μ restricted to the unstable leaves of $W^u(\Lambda_3(\mu))$, as explained in Chapter 4, Section 1.

For the next proposition and in Section 4, we will assume that φ_{μ_1, μ_2} preserves orientation. As observed at the end of the present section, for the purpose of proving the main theorem in Section 5 this is no restriction.

PROPOSITION 2. *For each $m \geq 3$ there is a neighbourhood W' of $(-2, 0)$ in the μ half plane such that on W' the maps P, Q_m and Λ_m (see proposition*

1) are defined and such that for all $\mu \in W'$, $W^u(P(\mu))$ and $W^s(Q_m(\mu))$ have a transverse intersection. Furthermore there is a curve γ in the μ half plane approaching the point $(-2, 0)$ such that for $\mu \in \gamma$, $W^u(P(\mu))$ and $W^s(P(\mu))$ have a homoclinic tangency while $W^u(Q_m(\mu))$ crosses $W^s(P(\mu))$.

PROOF: The first statement follows easily: if μ is near $(-2, 0)$, $W^u(P(\mu))$ is close to ℓ and $W^s(Q_m(\mu))$ is almost horizontal. As to the second statement we observe that for $\mu_2 = 0$, the boundary of the unstable manifold of $P(\mu)$ crosses the second pre-image of $\varphi_\mu^{-1}(P(\mu)) \cap \ell$ for $\mu_1 < -2$ as can be seen from Figure 6.4.

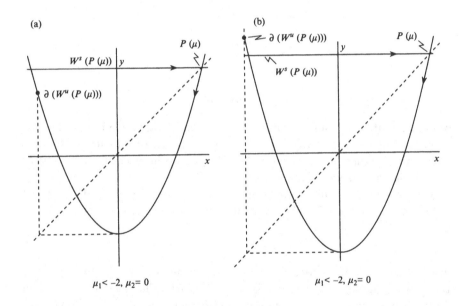

Figure 6.4

For $\mu_2 > 0$ but small and μ_1 increasing, we have a first tangency between $W^u(P(\mu))$ and $W^s(P(\mu))$ in the situation shown in Figure 6.5 (remember that φ_{μ_1, μ_2} preserves orientation).

Let γ be the curve consisting of these μ-values of first tangency (for fixed μ_2 and increasing μ_1 near -2). It clearly follows from the fact that $W^u(Q_m(\mu))$ and $W^u(P(\mu))$ cannot intersect, that $W^u(Q_m(\mu))$ crosses $W^s(P(\mu))$ for $\mu \in \gamma$.

Note that we did not (and need not) prove that this crossing of $W^u(Q_m(\mu))$ and $W^s(P(\mu))$ is transverse: in a later application which we make of the

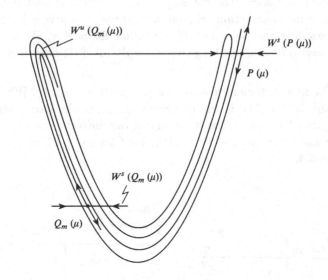

$W^u(Q_m(\mu))$

$W^s(P(\mu))$

$P(\mu)$

$W^s(Q_m(\mu))$

$Q_m(\mu)$

Figure 6.5

present result, it is enough that we can obtain a transverse intersection by an arbitrarily small perturbation. For this, the above result suffices. It turns out, however, that $W^u(Q_m(\mu))$ and $W^s(P(\mu))$ do intersect transversely, as will be shown in Appendix 4.

Now we come to the implications of our present results in the context of Section 4, Chapter 3. The rescaled one-parameter families of diffeomorphisms

$$\Psi_{n,\tilde{\mu}}^{-1} \circ \varphi_{M_n(\tilde{\mu})}^{n+N} \circ \Psi_{n,\tilde{\mu}}(\tilde{x}, \tilde{y})$$

introduced in that section, and from now on denoted by $\hat{\varphi}_{n,\tilde{\mu}}$, converge to the one-parameter family $\hat{\varphi}_{\tilde{\mu}}(\tilde{x}, \tilde{y}) = (\tilde{y}, \tilde{y}^2 + \tilde{\mu})$. This last one-parameter family is just the same as $\varphi_{\tilde{\mu},0}$ in the notation which we used up to now. The fact that $\hat{\varphi}_{n,\tilde{\mu}}$ converges to $\hat{\varphi}_{\tilde{\mu}}$ in the C^2 topology implies that our previous propositions are also applicable in the present case. Observe that $\hat{\varphi}_{n,\tilde{u}}$ may be assumed to preserve orientation, see the final remark in Chapter 3. So we have the following.

PROPOSITION. *For the one-parameter families of diffeomorphisms*

$$\hat{\varphi}_{n,\tilde{\mu}}(\tilde{x}, \tilde{y}) = \Psi_{n,\tilde{\mu}}^{-1} \circ \varphi_{M_n(\tilde{\mu})}^{n+N} \circ \Psi_{n,\tilde{\mu}}(\tilde{x}, \tilde{y})$$

as above, we have for $\tilde{\mu}$ near -2 and n sufficiently big

  – *there is a fixed saddle point $P_n(\tilde{\mu})$ near $(\tilde{x}, \tilde{y}) = (2, 2)$,*

- there is a point $Q_{n,m}(\tilde{\mu})$ of period m belonging to a basic set $\Lambda_m(\tilde{\mu})$ with stable thickness of $(\Lambda_m(\tilde{\mu}), Q_{n,m}(\tilde{\mu}))$ near $2^{m-1} - 3$ (up to a bounded factor; see the conjugacy between the tent and logistic maps),

- $W^s(Q_{n,m}(\tilde{\mu}))$ and $W^u(P_n(\tilde{\mu}))$ have a transverse intersection.

Furthermore, there are values γ_n, converging to -2, such that for $\tilde{\mu} = \gamma_n$, $W^u(P_n(\tilde{\mu}))$ and $W^s(P_n(\tilde{\mu}))$ have a homoclinic tangency while $W^u(Q_{n,m}(\tilde{\mu}))$ is crossing $W^s(P_n(\tilde{\mu}))$.

§4 Separatrices of saddle points for diffeomorphisms near a homoclinic tangency

We return to a one-parameter family of two–dimensional diffeomorphisms φ_μ as in Section 4 of Chapter 3, i.e. with a homoclinic tangency for $\mu = 0$. It yields a sequence of one-parameter families of diffeomorphisms

$$\hat{\varphi}_{n,\tilde{\mu}}(\tilde{x}, \tilde{y}) = \Psi_{n,\tilde{\mu}}^{-1} \circ \varphi_{M_n(\tilde{\mu})}^{n+N} \circ \Psi_{n,\tilde{\mu}}(\tilde{x}, \tilde{y})$$

which were again discussed in the previous section. There, for $\tilde{\mu}$ near -2 and n big, we denoted by $P_n(\tilde{\mu})$ the fixed point of $\hat{\varphi}_{n,\tilde{\mu}}$ which is close to $(\tilde{x}, \tilde{y}) = (2, 2)$. Corresponding to this fixed point, there is a periodic point with period $n + N$ of φ_μ which we also denote by $P_n(\mu)$. (When $\mu, \tilde{\mu}$, and n appear simultaneously, it is understood that they are related by $\mu = M_n(\tilde{\mu})$, see the theorem in Section 4, Chapter 3). In this section, which is entirely dedicated to the proof of the next propostion, we analyse the stable and unstable separatrices of $P_n(\mu)$ for μ near $M_n(-2)$.

PROPOSITION 1. *For $\tilde{\mu}$ near -2, there are compact arcs $\sigma_n^s(\tilde{\mu})$ and $\sigma_n^u(\tilde{\mu})$ in $W^s(P_n(\tilde{\mu}))$ and $W^u(P_n(\tilde{\mu}))$ containing $P_n(\tilde{\mu})$ and converging, for $n \to \infty$, to an arc in $W^s(p_0)$, respectively in $W^u(p_0)$, containing at least one fundamental domain (p_0 is the saddle point of φ_0 related to the homoclinic tangency).*

PROOF: In this proof we shall have to work in three different scales which we describe first.

- The scale of the $\tilde{x}, \tilde{y}$ coordinates defined by $\Psi_{n,\tilde{\mu}}(\tilde{x}, \tilde{y}) = (x, y)$. This scale decreases for increasing n: for each $(\tilde{x}, \tilde{y})$, $\lim_{n \to \infty} \Psi_{n,\tilde{\mu}}(\tilde{x}, \tilde{y}) = q$, a point of the orbit of tangency. In Figure 6.6 we indicate the size and position of a unit square in the $(\tilde{x}, \tilde{y})$-coordinates as function of n.

- The scale of a small but fixed neighbourhood U of the point q of tangency of $W^u(p_0)$ and $W^s(p_0)$. Before choosing U we construct stable and unstable foliations $\mathcal{F}_\mu^s$ and $\mathcal{F}_\mu^u$ in a neighbourhood of p_μ – see also the last remark of

Section 4, Chapter 3. We assume that these foliations are C^1 and depend continuously on μ. One may, but need not, derive them from the linearizing coordinates. Iterating $\mathcal{F}_\mu^u$ by φ_μ, respectively $\mathcal{F}_\mu^s$ by φ_μ^{-1}, we can obtain that they are both defined near q and have a line of tangencies, which we denote by ℓ_μ. We choose U so that for μ near zero and any leaf F of $\mathcal{F}_\mu^u$ or $\mathcal{F}_\mu^s$, the angle between its tangent and the tangent to the other foliation is strictly increasing along F in the direction away from $F \cap \ell_\mu$. In Figure 6.7 we show $W^s(p_\mu), W^u(p_\mu)$, the foliations $\mathcal{F}_\mu^s$ and $\mathcal{F}_\mu^u$ and the line of tangencies ℓ_μ, for μ near zero, in a neighbourhood of q.

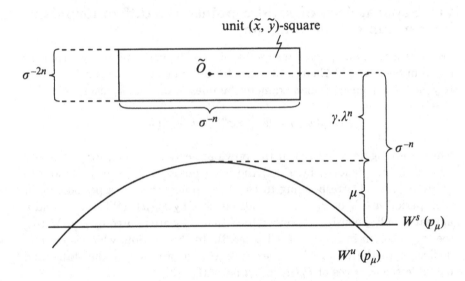

Figure 6.6

(*In Figure 6.6, as in Chapter III, Section 4, σ is the expanding and λ the contracting eigenvalue of $(d\varphi)_{P_\mu}$; γ is independent of n; $(\tilde{x} = \tilde{y} = 0)$ is denoted by $(0,0)$.*)

– Finally there is the *global* scale: the scale of fundamental domains of $W^u(p_0)$ and $W^s(p_0)$.

On the scale of the $(\tilde{x}, \tilde{y})$-coordinates we know the stable and the unstable manifold of $P_n(\tilde{\mu})$, or at least their limits for $n \to \infty$. We shall only consider one branch in each of these separatrices as indicated in Figure 6.8 (for $\tilde{\mu}$ near -2 and n big).

As observed before, the diffeomorphisms $\hat{\varphi}_{n,\tilde{\mu}}$ are converging to $\hat{\varphi}_{\tilde{\mu}}(\tilde{x}, \tilde{y}) = (\tilde{y}, \tilde{y}^2 + \tilde{\mu})$ and the foliations $\mathcal{F}_\mu^s$ and $\mathcal{F}_\mu^u$, expressed in the $(\tilde{x}, \tilde{y})$-coordinates, are converging for $n \to \infty$ to the horizontal foliation and the foliation by the parabolas $\{\tilde{y} = \tilde{x}^2 + a \mid a \in \mathbb{R}\}$. This convergence is uniform in the C^2-topology on compact parts of the $(\tilde{x}, \tilde{y})$–plane. So, for (big) $K > 0$ we

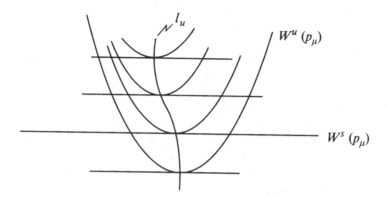

Figure 6.7

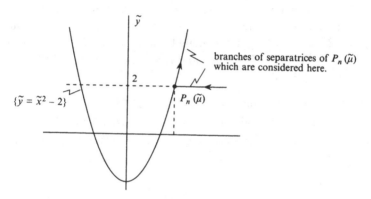

Figure 6.8

have for n sufficiently big, say $n \geq n(K)$, compact arcs $\hat{\sigma}_n^u(\tilde{\mu})$ and $\hat{\sigma}_n^s(\tilde{\mu})$ in the branches of $W^u(P_n(\tilde{\mu}))$ and $W^s(P_n(\tilde{\mu}))$ indicated in Figure 6.8 (and containing $P_n(\tilde{\mu})$) such that

- the angles between leaves of $\mathcal{F}_\mu^s$ and the *last fundamental domain of* $\hat{\sigma}_n^u(\tilde{\mu})$, i.e. $\hat{\sigma}_n^u(\tilde{\mu}) - \hat{\varphi}_{n,\tilde{\mu}}^{-1}(\hat{\sigma}_n^u(\tilde{\mu}))$, are at least arctan (K),

- the angles between $\mathcal{F}_\mu^u$ and $\hat{\sigma}_n^s(\tilde{\mu})$ are at least $\frac{\pi}{4}$,

- the angles between leaves of $\mathcal{F}_\mu^s$ and $\hat{\sigma}_n^s(\tilde{\mu})$ and between leaves of $\mathcal{F}_\mu^u$ and $\hat{\sigma}_n^u(\tilde{\mu})$ are at most K^{-1}.

All these angles are taken with respect to the $(\tilde{x}, \tilde{y})$-coordinates. In order to get these angles with respect to the (x, y)-coordinates we have to replace each angle α by arctan $(\sigma^{-n} \cdot \tan(\alpha))$, see Figure 6.6 illustrating the $(\tilde{x}, \tilde{y})$-

scale. This means that we get small angles (for n big), so that we can neglect the difference between an angle λ and $\tan(\lambda)$ and obtain

- between $\mathcal{F}_\mu^s$ and $\hat{\sigma}_n^s(\tilde{\mu})$ and between $\mathcal{F}_\mu^u$ and $\hat{\sigma}_n^u(\tilde{\mu})$, at most $K^{-1} \cdot \sigma^{-n}$,

- between $\mathcal{F}_\mu^s$ and the last fundamental domain of $\hat{\sigma}_n^u(\tilde{\mu})$, at least $K \cdot \sigma^{-n}$,

- between $\mathcal{F}_\mu^u$ and $\hat{\sigma}_n^s(\tilde{\mu})$, at least σ^{-n}.

Next we extend $\hat{\sigma}_n^u(\tilde{\mu})$ using iterations of φ_μ^{n+N} (with $\mu = M_n(\tilde{\mu})$) and extend $\hat{\sigma}_n^s(\tilde{\mu})$ using iterations of $\varphi_\mu^{-(n+N)}$ and show that, as long as these iterations are in U, the corresponding arcs, which are denoted by $\hat{\sigma}_n^u(\tilde{\mu})$ and $\hat{\sigma}_n^s(\tilde{\mu})$, are very close to leaves of $\mathcal{F}_\mu^u$ and $\mathcal{F}_\mu^s$. For this we need the following two claims, in which C is a constant depending only on $\varphi_\mu^N \mid \varphi_\mu^{-N}(U)$ (the diffeomorphism from a neighbourhood of $(x, y) = (0, 1)$ to the neighbourhood U of $(x, y) = (1, 0) = q$, where (x, y) are the linearizing coordinates around p_μ) but not on n.

CLAIM 1: For n sufficiently big and $v \in T_w M$, with $w \in U$ such that $\varphi_\mu^i(w)$ is in the domain of the linearizing coordinates for $i = 0, \ldots, n$, such that $\varphi_\mu^{n+N}(w) \in U$, and such that the angle between v and $\mathcal{F}_\mu^s$ is at least $K \cdot \sigma^{-n}$, we have $\|d\varphi_\mu^{n+N}(v)\| \geq C \cdot K \cdot \|v\|$ and the angle between $d\varphi_\mu^{n+N}(v)$ and $\mathcal{F}_\mu^u$ is at most $C \cdot K^{-1} \cdot \lambda^n$.

CLAIM 2: For n sufficiently big and $v \in T_w M$, with $w \in U$ such that $\varphi_\mu^{-(N+i)}(w)$ is in the domain of the linearizing coordinate for $i = 0, \ldots, n$, such that $\varphi_\mu^{-(n+N)}(\omega) \in U$ and the angle between v and $\mathcal{F}_\mu^u$ is at least σ^{-n}, we have $\|d\varphi_\mu^{-(n+N)}(v)\| \geq C \cdot (\sigma \cdot \lambda)^{-n} \cdot \|v\|$ and the angle between $d\varphi_\mu^{-(n+N)}(v)$ and $\mathcal{F}_\mu^s$ is at most $C \cdot \lambda^n$.

PROOF OF THE CLAIMS: Since both claims are proved essentially in the same way, we only show the details of the proof of the first claim. Let $v = v_x \frac{\partial}{\partial x} + v_y \frac{\partial}{\partial y}$ be a unit vector (with respect to the linearizing coordinates) as in the hypothesis of Claim 1. So $|v_y| \geq K \cdot \sigma^{-n}$ (since n is big we neglect the difference between $K \cdot \sigma^{-n}$ and $\sin(K \cdot \sigma^{-n})$ or $\tan(K \cdot \sigma^{-n})$). Then $d\varphi_\mu^n(v) = \lambda^n v_x \frac{\partial}{\partial x} + \sigma^n v_y \frac{\partial}{\partial y}$ which has length at least $\sigma^n \cdot |v_y| \geq K$. The angle between $d\varphi_\mu^n(v)$ and $\frac{\partial}{\partial y}$, the $\mathcal{F}_\mu^u$-direction, is at most $(\lambda^n \cdot |v_x|) \cdot (\sigma^n \cdot |v_y|)^{-1} \leq \lambda^n \cdot K^{-1}$. Choosing the constant C so that it takes into account the distortion due to $\varphi_\mu^N \mid \varphi_\mu^{-N}(U)$, we get the inequalities as stated in the conclusion of Claim 1. □

Indeed it follows from these two claims that, provided we choose K so big that $C \cdot K > 1$, the arcs $\hat{\sigma}_n^u(\tilde{\mu})$ and $\hat{\sigma}_n^s(\tilde{\mu})$ are very close to leaves of $\mathcal{F}_\mu^u$ and $\mathcal{F}_\mu^s$ and they are large enough to reach the boundary of U. This can be seen by observing that positive iterates of $\hat{\sigma}_n(\tilde{\mu})$ and negative iterates of $\hat{\sigma}_n^s(\tilde{\mu})$ move away from the line of tangencies, so that the claims may be applied repeatedly to their points and tangent vectors. Therefore, these arcs are close to $W^u(p_\mu)$ and $W^s(p_\mu)$ if n is big, and they have a limit for $n \to \infty$, namely an interval in $W^u(p_0)$, resp. $W^s(p_0)$, from the point q of tangency to the boundary of U. So for n big, $\varphi_\mu^N(\hat{\sigma}_n^u(\mu))$ and $\varphi_\mu^{-N}(\hat{\sigma}_n^s(\mu))$ project along the leaves of $\mathcal{F}_\mu^s$ and $\mathcal{F}_\mu^u$ onto segments in $W^u(p_\mu)$ and $W^s(p_\mu)$ that cover several fundamental domains. It follows that $\varphi_\mu^{(n+N)}(\hat{\sigma}_n^u(\tilde{\mu}))$ and $\varphi_\mu^{-(n+N)}(\hat{\sigma}_n^s(\tilde{\mu}))$ contain segments close to arcs in $W^u(p_\mu)$ and $W^s(p_\mu)$ that cover several fundamental domains. $\qquad\square$

REMARK 1: From the construction, we conclude that the fundamental domains which are "pursued" by the arcs $\varphi_\mu^{(n+N)}(\hat{\sigma}_n^u(\tilde{\mu}))$ and $\varphi_\mu^{-(n+N)}(\hat{\sigma}_n^s(\tilde{\mu}))$ belong to the same branches of $W^u(p_\mu)$ and $W^s(p_\mu)$ as the homoclinic tangency we started with. Also, as we have seen in Chapter 3, we may assume that these same branches are involved in a nontrivial basic set. This implies that these arcs $\varphi_\mu^{(n+N)}(\hat{\sigma}_n^u(\tilde{\mu}))$ and $\varphi_\mu^{-(n+N)}(\hat{\sigma}_n^s(\tilde{\mu}))$ intersect $W^s(p_\mu)$ and $W^u(p_\mu)$ transversally for n sufficienty big.

§5   Proof of the main result

In this section we combine all the previous material to prove the following result.

THEOREM 1. *Let $\varphi \in \mathrm{Diff}^2(M)$, M a 2-manifold, be a diffeomorphism with a saddle point p whose stable and unstable manifolds have an orbit of tangency. Then arbitrarily near φ there is an open set $U \subset \mathrm{Diff}^2(M)$ with persistent homoclinic tangencies. If moreover $|\det(d\varphi)_p| < 1$, then there is a residual subset $R \subset U$ such that each $\varphi' \in R$ has infinitely many hyperbolic periodic attractors (sinks).*

PROOF: We show that after four perturbations of φ, to be described below, we get a diffeomorphism which is in the boundary of an open set $U \subset \mathrm{Diff}^2(M)$ as required. The perturbations can be taken arbitrarily small. In order not to overload the notation we shall also denote the perturbed diffeomorphisms by φ. Below we describe what properties the sucessive perturbations of φ have; we can and will take these successive perturbations in such a way that the properties we obtained already for earlier perturbations remain (except where we explicitly state the opposite).

1. Since, by Section 1 of Chapter 3, homoclinic tangencies, in generic one-parameter families, are always accumulated by such tangencies, we can perturb φ so that it has, besides a homoclinic tangency, also transverse homoclinic intersections of $W^u(p)$ and $W^s(p)$. By Section 5, Chapter 2, the

perturbed φ has then a basic set which we denote by Λ_1. Also, Q_1 denotes a periodic (or fixed) point in Λ_1. So, stable and unstable manifolds of p and Q_1 have mutual transverse intersections, i.e. both $W^u(p) \cap W^s(Q_1)$ and $W^u(Q_1) \cap W^s(p)$ contain points of transverse intersection.

2. In the second perturbation, we choose φ to be C^∞, $|\det(d\varphi)_p| \neq 1$, and the eigenvalues of $(d\varphi)_p$ to satisfy the conditions for C^2 linearizability as in Section 4 of Chapter 3. Also, we make the contact of $W^u(p)$ and $W^s(p)$ at the orbit of tangency nondegenerate, i.e. parabolic. From now on we assume that $|\det(d\varphi)_p| < 1$ (otherwise one can proceed with φ^{-1} instead of φ). This means that the results of Section 4, Chapter 3, are applicable to any one-parameter family of C^∞ diffeomorphisms φ_μ if $\varphi_0 = \varphi$ and φ_μ unfolds the tangency generically.

3. The third perturbation is justified by the last proposition of Section 3 of the present chapter, which we can apply since φ embeds in a one-parameter family φ_μ as indicated above. After this perturbation, $W^u(p)$ and $W^s(p)$ do not need to have tangencies any more, but φ has

   – a periodic point P,

   – a basic set Λ_2 with periodic point Q_2

(in the proposition P, Λ_2, and Q_2 were denoted by $P_n(\tilde{\mu})$, $\Lambda_m(\tilde{\mu})$, and $Q_{n,m}(\tilde{\mu})$, with $\tilde{\mu} = \gamma_n$) such that the stable thickness of (Λ_2, Q_2) is big (in particular we assume that $\tau^s(\Lambda_2, Q_2) \cdot \tau^u(\Lambda_1, Q_1) > 1$), $W^s(Q_2)$ and $W^u(P)$ have a transverse intersection, $W^u(P)$ and $W^s(P)$ have a homoclinic tangency, and $W^u(Q_2)$ crosses $W^s(P)$. Because of the results of the previous section we may assume that $W^u(P)$ and $W^s(P)$ are close to $W^u(p)$ and $W^s(p)$ over a distance of at least one fundamental domain. This means that the stable and unstable manifolds of Q_1 and P have mutually transverse intersections.

4. In the last perturbation we produce a transverse intersection of $W^u(Q_2)$ and $W^s(P)$.

Observe that for this last diffeomorphism, P, Q_1, and Q_2 all have mutually transversally intersecting stable and unstable manifolds while P has also a homoclinic tangency. Using $\tau^s(\Lambda_2, Q_2) \cdot \tau^u(\Lambda_1, Q_1) > 1$ and arguing as in the proofs of the corollaries in Section 1 of this chapter, we obtain after this last perturbation a diffeomorphism which is in the boundary of an open set $U \subset \mathrm{Diff}^2(M)$ with persistent homoclinic tangencies, as required.   □

REMARK 1: As a consequence of the main result in this chapter, we have that the diffeomorphisms with a hyperbolic limit set are not C^2-dense in the space of all C^2 surface diffeomorphisms. (Similarly, the L-stable or Ω-stable diffeomorphisms are not C^2-dense). This fact, however is still not known in the C^1 topology, so we pose the following

PROBLEM 1: Are the diffeomorphisms with a hyperbolic limit set C^1-dense in the space of all surface diffeomorphisms?

§6   Sensitive chaotic orbits near a homoclinic tangency

In this final section we want to discuss the results that we have obtained so far in terms of occurrence of sensitive orbits and strange attractors.

First, a homoclinic tangency, and its unfolding, give rise to hyperbolic basic sets of saddle type; in these basic sets, most orbits are sensitive. On the other hand, for a diffeomorphism in the plane near a homoclinic tangency, associated to a dissipative (area contracting) saddle point, there can be no nontrivial hyperbolic attractor near the orbit of tangency: such an attractor contains "holes" in its basin where the map must be expanding, see Plykin [**P**,1974], which is impossible in the dissipative case. So, as long as the dynamics is hyperbolic, the chaotic orbits can only occupy a set of Lebesgue measure zero.

Secondly, there is the phenomenon of the coexistence of infinitely many periodic attractors (sinks). Generically, the number of periodic attractors, with period smaller than some constant, is finite. Thus, if there are infinitely many attractors, most should have very big period. Of course, periodic attractors are not sensitive, but in numerical experiments, where one can analyse only a finite part of an orbit, these periodic attractors of very high period may look chaotic.

Finally, although we cannot expect nontrivial hyperbolic attractors, *we do expect non-hyperbolic strange attractors*. This is based on the fact that in the unfolding of a generic homoclinic tangency there are Hénon-like families of diffeomorphisms and for such families Mora and Viana [**MV**,1991] proved the existence of "persistent strange attractors". See the next chapter.

CHAPTER 7

OVERVIEW, CONJECTURES AND PROBLEMS –
A THEORY OF HOMOCLINIC BIFURCATIONS –
STRANGE ATTRACTORS

Based on recent developments, conveyed in the previous chapters and further discussed here, we now present some perspective, and indeed a programme, concerning homoclinic bifurcations and their relations to chaotic dynamics. Actually, we consider homoclinic bifurcations as a main mechanism to unleash a string of complicated changes in the dynamics of a diffeomorphism (or, more generally, an endomorphism).

Indeed, as we have seen, the one-parameter unfolding of a homoclinic tangency yields a striking number of dynamical phenomena:

- hyperbolic Cantor sets (Chapter 2),
- cascades of homoclinic tangencies (Chapter 3),

and for locally dissipative surface diffeomorphisms ·

- cascades of period doubling bifurcations (Chapter 3),
- relative prevalence of hyperbolicity of the limit set in a significant number of cases (Chapter 5, Appendix 5) and its converse (Chapter 7),
- infinitely many sinks (Chapter 6, Appendix 4),
- Hénon-like strange attractors, as proved in [**MV**,1991] extending the work in [**BC**,1991], to be discussed in this chapter.

Thus, homoclinic bifurcations embody most of the known bifurcations of a nonlocal character, at least in the setting of surface diffeomorphisms or three-dimensional flows without singularities. On the other hand, there is some evidence, still quite limited, that a nonhyperbolic diffeomorphism exhibiting (one of) the above complicated phenomena might be approximated by one exhibiting a homoclinic bifurcation. These considerations lead us to propose a number of related questions that when put together point toward a theory of homoclinic bifurcations.

Our programme, as it will be discussed in this chapter, consists of

(1) to determine a dense subset $\mathcal{H}$ of all dynamical systems in the C^k topology such that if $\varphi \in \mathcal{H}$, then either

  - φ is persistently hyperbolic (i.e., it satisfies Axiom A and has no cycles)

or

  - φ exhibits a bifurcation of some specified type,

(2) to unfold the elements in $\mathcal{H}$, establishing which dynamical phenomena

are more prevalent in terms of the parameter space: hyperbolicity of the limit set, strange attractors, infinitely many sinks, others (?).

Thus, the main objective is to capture "most" or "a large" part of the global complexity of dynamics beyond hyperbolicity. But, differently from previous programs with similar objectives, we do not suggest the description of the dynamics of the diffeomorphisms in an *open and dense or even a residual subset* (countable intersection of open and dense subsets) of $\text{Diff}^k(M)$: we ask which kinds of dynamics are, in a parametrized form, more common near the elements of a dense subset of $\text{Diff}^k(M)$.

In fact, we conjecture that the bifurcations of *specified type* referred to above should be homoclinic bifurcations: *homoclinic tangencies or (finite) cycles of periodic saddles of different index*. This is justified, in particular, by the dynamic richness that the unfolding of a homoclinic tangency yields, as we have just listed – see also Conjectures 1 and 2 in Section 1 below and the remarks following them.

Before going into more details, some remarks about the setting of our proposed programme are in order. First, some of the questions posed here can be adapted to one-dimensional noninvertible dynamics, as we shall briefly remark later. Secondly, a number of two-dimensional results concerning unfoldings of homoclinic tangencies are now also available in higher dimensions; see Section 4 in Chapter 5 and the introduction to Chapter 6 and also comments in the last section of this chapter. It seems to us, however, that more facts should be proved or discovered in order to begin to set forth a similar overall perspective in higher dimensions.

§1 Homoclinic bifurcations and nonhyperbolic dynamics

In this section we discuss how common it is for a nonhyperbolic diffeomorphism to exhibit homoclinic tangencies or, more generally, to be a homoclinically bifurcating map.

Let us first present some definitions. We say that a diffeomorphism is hyperbolic if its nonwandering set (or even its limit set) is hyperbolic. A weaker, but still very meaningful definition is to require the existence of a finite number of *hyperbolic attractors* whose basins (stable sets or realm of attraction) cover a subset of total Lebesgue measure in the ambient manifold. In this case, we say that the diffeomorphism is *essentially hyperbolic*.

A second main concept that we need is that of a homoclinic bifurcation. Roughly speaking, it means the creation of "new" (transverse) homoclinic orbits through small perturbations of the map. A particularly important case is that of a homoclinic tangency. There are, however, interesting examples, discussed in the sequel, of homoclinic bifurcations that do not correspond to homoclinic tangencies. Indeed, in some of these examples, in three-dimensions, there is a creation of transverse homoclinic orbits but no homoclinic tangency is exhibited by the initial diffeomorphism or any small perturbation of it. Therefore, a formal definition requires more care. We say that φ is homoclinically stable if for each continuous curve $(\varphi(t), x(t))$ in $\text{Diff}^k(M) \times M$ such that $\lim_{t \to t_0} \varphi(t) = \varphi$ and such that each $x(t)$ is a transverse homoclinic orbit of $\varphi(t)$, the limit $\lim_{t \to t_0} x(t)$ exists, and is a transverse homoclinic orbit of φ. If φ is not homoclinically stable, then we say by *abus de langage* that φ exhibits a *homoclinic bifurcation* or that φ is a *homoclinically bifurcating* diffeomorphism.

CONJECTURE 1. *Every $\varphi \in \text{Diff}^k(M)$, $k \geq 1$, can be approximated by a diffeomorphism which is (essentially) hyperbolic or else by one exhibiting a homoclinic bifurcation. Moreover, among the homoclinically bifurcating diffeomorphisms the ones exhibiting a homoclinic tangency or a (finite) cycle of hyperbolic periodic saddles of different indices form a dense subset.*

An especially interesting case is that of

CONJECTURE 2. *If $\dim M = 2$, then every $\varphi \in \text{Diff}^k(M)$ can be approximated by a diffeomorphism which is (essentially) hyperbolic or by one exhibiting a homoclinic tangency.*

REMARK 1: Concerning the above conjectures, we note that one can, of course, *enlarge* the set of bifurcating diffeomorphisms in $\mathcal{H}$ to include a few other relevant cases like, for example, diffeomorphisms with a Feigenbaum attractor (discovered independently by Coullet and Tresser), see [**CE**,1980] and [**GST**,1989]: these diffeomorphisms are accumulation points of sequences of diffeomorphisms exhibiting successive period doubling bifurcations of sinks of periods $p \cdot 2^n$, for some integer p and $n \in \mathbf{N}$. Also of much importance are the saddle-node critical cycles studied in [**NPT**,1983]: *their unfoldings yield homoclinic tangencies and vice-versa*. Moreover, by a recent work of Diaz, Rocha and Viana [**DRV**,1991], they correspond to points of positive denseness of Hénon-like strange attractors in their one-parameter unfoldings. See below and the last section of this chapter.

REMARK 2: We also observe that these conjectures are probably very hard to prove or disprove. But, they may very well be true in relevant open sets in $\text{Diff}^k(M)$. For instance, the second conjecture is obviously true in the open set of diffeomophisms which are persistently hyperbolic as well as the ones with persistent tangencies that we have constructed in Chapter 6. More encouraging, it has been shown to be true in general for C^2 diffeomorphisms but only in the C^1 topology [**AM**,1991]. *A good programme in this context is to show that diffeomorphisms exhibiting a main dynamic bifurcation like* a *Feigenbaum attractor, a Hénon-like attractor (see definition in the next section) or infinitely many coexisting sinks can be approximated by one with a homoclinic tangency or (see the previous remark) a critical saddle-node cycle.* This is the content of Problem 9 in Section 3.

Let us a make a brief comment about one-dimensional dynamics. Consider endomorphisms of the circle or the interval with all critical points being nondegenerate. Suppose φ is such an endomorphism with some critical point and let p be a periodic point (periodic points must exist by a result in [**BF**,1973]); we assume that p is hyperbolic and expanding. We say that φ exhibits a homoclinic tangency associated to p if, for some $x \in W_{\text{loc}}^u(p)$, there are $m, n \in \mathbf{N}$ such that $\varphi^m(x) = c$ and $\varphi^n c = p$, for some critical point c. In this setting Conjecture 2 above could be translated into the following one: *every endomorphism can be approximated either by a hyperbolic one or by one with a homoclinic tangency.* This statement in turn implies the main conjecture in this subject: the hyperbolic endomorphisms are C^k-dense in the space of all C^k endomorphisms for $k \geq 2$ (for $k = 1$ this has been proved in [**J**,1971]). But, we conjecture that even more is true: *every nonhyperbolic endomorphism can be approximated by one exhibiting a homoclinic tangency.* The idea is that if we look beyond purely local bifurcations (saddle-nodes and period doubling), then endomorphisms of the circle or the interval exhibiting homoclinic tangencies might be dense in the complement of the hyperbolic ones. Notice that a map with a nontrivial rotation interval, which changes under small perturbations of the map, can be C^∞ approximated by one with a homoclinic tangency [**NPT**,1983]. For rational maps of the Riemann sphere, homoclinic tangencies correspond to pre-periodic critical orbits. The conjecture here is that the endomorphisms with pre-periodic critical orbits are dense in the complement of the hyperbolic ones.[1]

Finally, let us mention that Conjecture 1 also makes sense for flows without singularities, by the usual considerations involving suspensions of diffeomorphisms (see e.g. [**PM**,1982]); similarly, Conjecture 2 applies to three-dimensional flows without singularities. However, to have a more global perspective, one has to take into account homoclinic bifurcations of singularities, like in Silnikov [**S**,1965], and the Lorenz-like attractors, which we will briefly discuss in the next section, as well as singular cycles (i.e., cycles involving a singularity and periodic orbits) studied in [**LP**,1986] and [**BLMP**,1991]. Without being too precise, we believe that some form of Conjectures 1 and 2 may still be formulated: the dense set $\mathcal{H}$, referred to in the conjectures for diffeomorphisms, should in this case also include

---

[1] Added in proof: Swiatek has just announced that the conjectures in this paragraph are true for real quadratic maps of the interval. He makes use of previous partial results of Yoccoz (also valid in the complex case) and Sullivan, Jacobson and himself, among others.

flows exhibiting homoclinic bifurcations of singularities, Lorenz-like attractors and singular cycles, besides the hyperbolic ones and the ones exhibiting
homoclinic bifurcations of periodic orbits.

We now discuss a number of interesting examples related to the above
conjectures. First of all, homoclinic orbits do not occur for diffeomorphisms
of the circle, except if we consider the case of a unique periodic point which
is a saddle-node. Anyhow, each $\varphi \in \mathrm{Diff}^k(S^1)$ can be approximated by a
hyperbolic (Morse–Smale) element. The bifurcating diffeomorphisms from
Anosov to DA maps ([**S**,1967], [**W**,1970]) exhibit homoclinic bifurcations but
cannot be approximated by diffeomorphisms with homoclinic tangencies;
however, they can be approximated by an Anosov or a DA diffeomorphism
which is, of course, hyperbolic.

Another important example is that of a surface diffeomorphism exhibiting
a saddle-node critical cycle; see Figure 7.1 where, in Figure 7.1(a), a critical
1-cycle is shown.

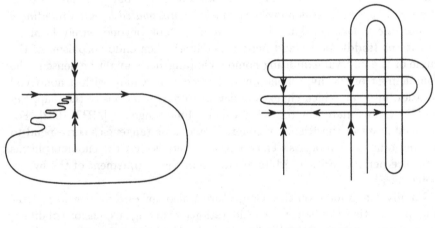

Figure 7.1(a)   Figure 7.1(b)

The 1-cycle case is obtained by collapsing a sink and a saddle (see Chapter
3), the unstable separatrix being in the basin of the saddle-node and having
orbits of tangencies with leaves of the strong stable foliation. In general, for
k-cycles, $k > 1$, there are $k - 1$ (hyperbolic) saddles taking part in the cycle
(see [**NPT**,1983]) and we can even consider the case of a sink collapsing
with a saddle whose stable and unstable manifolds bound a horseshoe; see
Figure 7.1(b). Here, we can approximate the bifurcating diffeomorphism by
a hyperbolic one before the collapsing or, more interestingly, by a diffeomorphism exhibiting a homoclinic tangency associated with a periodic saddle
of very high period that appears after the unfolding of the saddle-node.

In higher dimensions there are very interesting examples of open sets of *nonhyperbolic diffeomorphisms: the elements exhibit no homoclinic tangencies, but they do exhibit homoclinic bifurcations.* One set of examples is due to Diaz [**D**,1991]: it is obtained after unfolding a three-dimensional cycle with two fixed saddles p and q whose stable indices are 1 and 2, respectively. We must have a transversal intersection of $W^u(p)$ and $W^s(q)$ as indicated in Figure 7.2. The cycle is then completed with a nontransversal orbit of intersection between $W^u(q)$ and $W^s(p)$ that we unfold. We also have to assume certain relations involving the eigenvalues of the map at p and q as well as other properties of the map along the cycle. Notice that the figure indicates that the intersection between $W^u(p)$ and $W^s(q)$ is connected; see [**DR**,1991] for the case where this intersection is not connected.

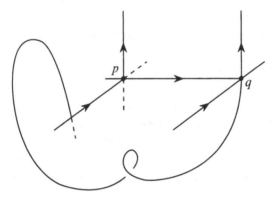

Figure 7.2

Another relevant set of examples is provided by Shub [**S**,1971] and Mañé [**M**,1978] as well as by Carvalho [**C**,1991]. In the case considered by Mañé there is an open set $\mathcal{V} \subset \text{Diff}^1(T^3)$, T^3 being the 3-torus, such that for $\varphi \in \mathcal{V}$ we have

- φ has two saddles of different indices whose stable and unstable manifolds are dense in T^3,

- there is a $d\varphi$-invariant decomposition, $TT^3 = E^s \oplus E^u \oplus E^c$ such that E^s is $d\varphi-$contracting and E^u is $d\varphi-$expanding,

- there are C^0 foliations with C^1 leaves $\mathcal{F}^1, \mathcal{F}^u$ and $\mathcal{F}^c$ tangent to E^s, E^u and E^c, respectively,

- $\mathcal{F}^c$ is normally and plaque expansive (see Appendix 1 and [**HPS**,1977]).

Then all of T^3 is a nonhyperbolic attractor for φ. On the other hand, these diffeomorphisms are homoclinically unstable or, according to our terminology, they exhibit homoclinic bifurcations. Indeed, for a dense subset of $\tilde{\varphi}$'s in $\mathcal{V}$, the one-dimensional invariant manifolds of the saddles intersect each other and these intersections are in the closure of (transversal) homoclinic points. On the other hand, for another dense subset of elements in $\mathcal{V}$, such manifolds should not intersect at all by the transversality theorem.

§2   Strange attractors

The notion of attractor is clearly of fundamental importance in dynamics since "many" points (positive measure, total measure) or large topological sets (open nonempty, open and dense) should have attractors as their ω-limit sets. Thus, a good dynamical understanding of attractors should explain what can be expected of the future behaviour of "most" orbits.

In order to include several interesting cases, we will present *more than one notion of attractor*, basically following Milnor's discussion on the subject [**M**,1985].

We say that A is an attractor for $\varphi: M \to M$ if

- *the basin of attraction (stable set, realm of attraction)*, consisting of points $x \in M$ for which $\omega(x) \subset A$, must have positive Lebesgue measure *or*, even sharper, it must be open nonempty (one may even require that A should be in the interior of its basin of attraction),

- there is no strictly smaller closed set $A' \subset A$ so that the basin of attraction of A' coincides with that of A up to a set of Lebesgue measure zero *or* φ is transitive on A.

We now give an initial presentation of relevant classes of attractors, to which we shall come back later to discuss their dynamical properties in more detail.

First, we notice that hyperbolic attractors satisfy at once all the above conditions in their sharpest form.

The same is true about the famous Lorenz attractors, at least for the geometric ones considered by Guckenheimer and Williams [**GW**,1979], [**W**,1979] and the third degree polynomial cases studied by Rychlik [**R**,1989] and Robinson [**R**,1989]. Lorenz's original and most remarkable work [**L**,1963] dealt numerically with the three-dimensional quadratic vector field

$$\dot{x} = 10x + 10y$$
$$\dot{y} = 28x - y - xz$$
$$\dot{z} = -\frac{8}{3}z + xy$$

that was obtained by truncation of a development in series of the Navier–Stokes equation associated to a *weather prediction* experiment. He "showed" the existence of an attractor containing the origin in $\mathbb{R}^3$ (a singularity or a zero of the vector field) and also infinitely many periodic orbits! Notice that there are two negative and one positive eigenvalues of the linear part of the vector field at the origin satisfying the relations $\lambda_1 < \lambda_2 < 0 < \lambda_3$, $\lambda_1 + \lambda_3 < 0$ and $\lambda_2 + \lambda_3 > 0$. In fact, we have a *strong contraction "normally" to the attractor* $(\lambda_1 < \lambda_2 < 0)$, and an *expanding behaviour along it* $(\lambda_2 + \lambda_3 > 0)$. Moreover, at random the orbits near the attractor seemed to be sensitive with respect to their initial positions in their future behaviour, as will be discussed shortly. From that *he concluded that no matter how precise are the present data one cannot predict the weather in the long run with total precision.* And all these properties seemed to be *robust*, i.e. they "remained valid" under small perturbations of the coefficients in the equations above; again, this will be discussed in the sequel. (See Figure 7.3(a)) Recently, a new and very interesting kind of Lorenz attractor, of a *contracting* nature along the attractor, unlike the previous ones, has been introduced by Rovella [**R**,1991]: at the singularity in the attractor, the linear part of the vector field has two negative eigenvalues whose norms are bigger than that of a positive one. Actually, this attractor had been presented by Arneodo, Coullet and Tresser [**ACT**,1981]. However, they focus their attention on the cascades of period doubling bifurcations and not on the subtle questions concerning the *persistence* of the attractor, as done by Rovella and discussed in the sequel; see Figure 7.3(b).

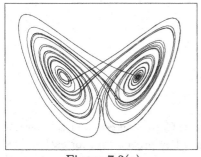

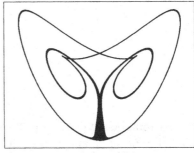

Figure 7.3(a)   Figure 7.3(b)

Also very interesting are the examples of Shub and Mañé of open sets of transitive *"partially hyperbolic"* diffeomorphisms on tori, mentioned in the last section and viewed here as attractors. Other nonhyperbolic DA-type attractors have been recently studied in [**C**,1991].

On the other hand, Bowen's example [**B**,1975b] of a C^1 (*but not C^k for $k \geq 2$!*) positive Lebesgue measure horseshoe, detailed in Chapter 4, is an attractor only from the measure-theoretical point of view.

A special role is played by the quasi-periodic systems on tori, which are conjugate to irrational translations. In the early work of Landau and Hopf, the quasi-periodic attractors were mentioned in their discussion on mathematical models for turbulence. Later, in [**RT**,1971], they were somewhat downplayed in that capacity as being *too simple or not sufficiently erratic.* However, they are *remarkably not so fragile (!) under perturbations even among nonconservative systems,* as we shall comment later. Thus, besides their fundamental role in conservative dynamics, they also do have an important place in the study of attractors for general systems.

In the context of this book, a recent important development has began with the work of Benedicks and Carleson concerning attractors for the two-dimensional dissipative (area contracting) Hénon family

$$\varphi_{a,b}(x,y) = (1 - ax^2 + y, bx).$$

Notably Hénon [**H**,1976] had asked if for $a = 1.4$ and $b = .3$ the corresponding map in the family exhibits a strange attractor, as his computational experiment seemed to indicate as a distinct possibility. See Figure 7.4.

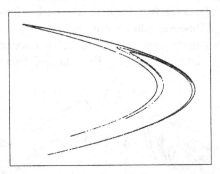

Figure 7.4

In their remarkable paper, Benedicks and Carlenson consider the parameters $b > 0$ to be small and a to be near 2. They proved that for each b fixed, there is a positive Lebesgue measure set of a-values E_b such that if $a \in E_b$, then $\varphi_{a,b}$ has an attractor $A_{a,b}$, which is the closure of the unstable manifold of some saddle $p_{a,b}$ and whose basin contains an open set, and moreover, that there is a point $z \in W^u(p_{a,b})$ such that its positive orbit is dense in $A_{a,b}$ and

$$\|(d\varphi_{a,b}^n)_z(v)\| \ge e^{cn} \|v\|$$

for all $n \ge 0$, some constant $c > 0$ and some vector $v \ne 0$. This shows that there are no sinks in the closure of $W^u(p_{a,b})$, i.e. $A_{a,b}$. By Plykin's work [**P**,1974], other hyperbolic attractors are also excluded, as already mentioned at the end of Chapter 6.

Subsequently, this result has been much extended by Mora–Viana [**MV**,1991]: the Hénon-like attractors as above appear whenever one generically unfolds a quadratic homoclinic tangency through one-parameter families of locally dissipative surface diffeomorphisms. Locally dissipative here means that the product of the eigenvalues at the saddle associated with the homoclinic tangency has norm less than 1. Actually, Mora and Viana also assumed that near such a saddle the map is C^3-linearizable. These generic conditions can be considerably weakened, for real analytic families, according to recent work by the second author of this book [**T**,1992].

In general in a manifold M of dimension 2 or more, we define *a nonhyperbolic attractor* A for a diffeomorphism φ to be *Hénon-like of codimension-1* if

- there is a saddle point $p \in A$ whose unstable manifold has dimension 1 and A is the closure of $W^u(p)$,

- the stable set (basin) of A contains an open set,

- there are a point $z \in A$ whose positive orbit is dense in A and a nonzero vector $v \in T_z M$ such that

$$\|d\varphi^n(v)\| \geq e^{cn} \|v\|$$

for all $n \geq 0$ and some constant $c > 0$.

Recently, Viana [**V**,1991b] has shown the existence of codimension-1 Hénon-like strange attractors extending to higher dimensions the previous results; see more comments at the end of Section 3.

Notice that in half the cases (namely, for orientation reversing diffeomorphisms) these Hénon-like attractors are in the interior of their basin of attraction.[2] We are then led to the following question.

PROBLEM 1: Are the Hénon-like attractors, that appear when unfolding homoclinic tangencies for orientation preserving surface diffeomorphisms, always in the interior of their basin of attraction?

Before this important work on the Hénon-like attractors, there was a pioneer result by Jacobson [**J**,1981] for the one-dimensional map $f_a(x) = 1 - ax^2$: for a in a positive Lebesgue measure subset of $\mathbb{R}$, f_a has an attractor carrying an absolutely continuous (with respect to Lebesgue) invariant measure. A few years later, similar results with somewhat different proofs were obtained by Rees [**R**,1986] in the context of rational mappings of the Riemann sphere and by Benedicks and Carleson [**BC**,1985] again for $f_a(x) = 1 - ax^2$.

---

[2]Added in proof: Viana has shown that this is true in all known cases. Problem 1 still makes sense for general Hénon-like attractors as defined above.

We now turn to the notion of *strange attractor*. Also here we present a number of alternative definitions. We say that an invariant set A for the diffeomorphism φ is a *strange attractor* if

- A is an attractor, as defined before, with basin $B(A)$,

- with total probability on $B(A) \times B(A)$, φ has sensitive dependence on initial conditions *or* almost all orbits (total probability) on $B(A)$ are sensitive.

Other properties are considered relevant concerning the *strangeness* of an attractor. For instance, to avoid very fragile dynamical structures, we can require a strange attractor to be *persistently strange* under small perturbations of the dynamical system: for every (generic) k-parameter family $\{\varphi_\mu\}$ of diffeomorphisms containing φ, there is a positive Lebesgue measure set $E \subset \mathbb{R}^k$ such that if $\mu \in E$ then φ_μ exhibits an attractor satisfying the above conditions for strangeness. When there is a strange attractor for every system near the initial one, we say that the attractor is *fully persistently strange*. Another property of relevance for the attractors is to have "*some degree*" of *exponential (geometric) sensitivity* of nearby orbits, i.e. future orbits of nearby points get apart at an exponential rate; this idea will be discussed later in this section. Let us see which of the attractors we presented before satisfy some of these notions.

We begin with the *nontrivial hyperbolic attractors:* they are *strange in all senses*. In fact, they are in the interior of their basin of attraction, they are *transitive* and also *exponentially sensitive*, i.e., up to the expansivity constant, almost all nearby points have their future orbits diverging at a geometric rate. Finally, they are *fully persistently strange* and even *structurally* or *dynamically stable*: all C^k-nearby systems have nearby nontrivial hyperbolic attractors which are dynamically equivalent to the initial one; see Chapter 1.

The *expanding Lorenz-like attractors*, i.e. the geometric ones and the third degree polynomial cases mentioned above, are also *strange in all senses*, perhaps even more so than the hyperbolic ones since they are dynamically unstable. They are in the interior of their basin of attraction, *transitive* and *exponentially sensitive*: the "canonical" one-dimension quotient of the Poincaré return map is expanding, i.e. its derivative has norm bigger than 1 in all points; see the previous references. They are also *fully persistently strange* but *dynamically unstable*: two real parameters are needed to describe the different equivalence classes of nearby systems or, in other words, they have *moduli of stability 2*. *Curiously enough we do not know whether these facts are mathematically true for the original vector field (or a small perturbation of it) provided by Lorenz thirty years ago!* This is of course a good problem to solve. Concerning the *contracting Lorenz-like attractors* exhibited by Rovella, the situation is more subtle: they are in the interior of their basin of attraction, *transitive* and *persistently strange* but *not fully* so. Also, it is very likely that they exhibit some degree of *exponential sensitivity*, as discussed later in the context of invariant measures and positive Lyapunov exponents. Finally, they are *dynamically unstable*.

The *"partially" hyperbolic attractors* exhibited by Shub, Mañé and, more recently, Carvalho are also *strange in all senses* and, like the expanding Lorenz attractors, they are *fully persistently strange* and *dynamically unstable*. Also, they are *exponentially sensitive*: this follows from the existence of a continuous and invariant unstable subbundle and the fact that the set of points in each centre-stable manifold has Lebesgue measure zero; finally, we observe that for any pair of points either they both belong to the same centre-stable manifold or their orbits will exponentially diverge in the future.

The *Hénon-like attractors*, shown to exist in the unfolding of homoclinic tangencies, are *transitive* and *persistently strange*, but most likely *not fully persistent* nor *dynamically stable*. Indeed this is the case for the Hénon-like attractors shown to exist in [**BC**,1991], [**MV**,1991] and [**V**,1991b]. This fact follows from a recent work of Ures [**U**,1992] stating that there are nearby diffeomorphisms exhibiting homoclinic tangencies and, by unfolding these tangencies, other nearby diffeomorphisms exhibiting sinks (attracting cycles) as originally asked in [**BC**,1991]. The Hénon-like attractors also have a good degree of *exponential sensitivity* as commented at the end of this section.

Let us now mention that the quasi-periodic attractors are not sensitive with respect to the initial conditions. They are, however, *persistent*, although *not fully so*. Originally, quasi-periodic orbits were studied in the context of Classical Mechanics. Later on, this have been extended to general dissipative (nonconservative) systems. According to our definition, the quasi-periodic attractors are *not strange* since they are *not sensitive*. However, if the number of frequencies (which is the dimension of the attractor) is big, then the dynamics looks quite complicated. Thus, they play a role between very simple (periodic) and *strange* attractors. About their celebrated persistence, we refer the reader to some classical and some recent literature: [**P**,1890], [**SM**,1971], [**K**,1957], [**A**,1961b], [**M**,1962], [**M**,1967], [**H**,1977], [**H**,1979], [**BHTB**,1990].

We close this section by describing, sometimes informally, how the idea of "some degree" of exponential dynamical behaviour or sensitivity has been pursued beyond the uniformly or even the partially hyperbolic diffeomorphisms that we have already discussed. For that, we first need to introduce various notions of entropy and the notion of Lyapunov exponent.

Let φ be a diffeomorphism of a compact manifold M, as before. The *entropy* $h(\varphi)$ of φ is a notion to quantify the partial predictability (predictability horizon): an initial state, specified up to precision $\varepsilon > 0$, leads, on average, after time $1/h(\varphi)$ to a state that can be predicted only up to precision 2ε. In the sequel we shall discuss the relation between this notion and that of sensitivity of the trajectories with respect to the initial conditions. Before that, we should make the notion more precise and one way to do this is to introduce the following concept of *topological entropy*.

We say that $S \subset M$ is an (n, ε)-generating set for φ, $n \in \mathbb{N}$ and $\varepsilon > 0$, if for every $x \in M$ there is $y \in S$ such that $d(\varphi^j(x), \varphi^j(y)) \leq \varepsilon$ for all $0 \leq j \leq n$. Since M is compact, it is easy to show that for every pair (n, ε) there is a finite generating set.

Thus, given (n, ε), we can define the "smallest number of points" $r(n)$ in the (n, ε)-generating sets: it represents the number of different orbits of length n that we "see" with precision ε. Then the topological entropy $h_{\text{top}}(\varphi)$ or just $h(\varphi)$ is defined by

$$h(\varphi) = \lim_{\varepsilon \to 0} \limsup_{n \to \infty} \frac{1}{n} \log r(n, \varepsilon).$$

Notice that all homeomorphisms of the circle have topological entropy zero. For Axiom A diffeomorphisms, the topological entropy is zero if and only if the nonwandering set is finite. If φ is a toral automorphism (linear endomorphism), then

$$h(\varphi) = \sum_{\lambda_i > 0} \lambda_i$$

where $\{e^{\lambda_i}\}$ is the set of eigenvalues of the linear map defining φ. For the "standard" horseshoe map $\psi \,|\, R$ (Figure 2.14(a)) or its variation (Figure 2.14(b)) the topological entropy is $\log 2$. Notice that if $\varphi^N = \psi$, after suitable restriction and for some integer $N > 0$, as in the previous sections of Chapter 2, then the contribution of this horseshoe to the topological entropy of φ is $\frac{1}{N} \log 2$.

A disadvantage of the above definition of topological entropy is that it does not distinguish between points whose neighbourhoods are "visited" more often by orbits (e.g. points in an attractor) from other points. This is very much taken into account when we consider a notion of entropy that depends on a Borel probability measure μ which is invariant under φ. A measure is called ergodic if every φ-invariant Borel set has measure either 1 or 0.

The definition of metric entropy $h_\mu(\varphi)$ can be formulated as follows. Given a partition $\mathcal{P} = \{P_1, \ldots, P_m\}$ of the support of μ, we first set

$$H_\mu(\mathcal{P}) = -\sum_{i=1}^{m} \mu(P_i) \log \mu(P_i)$$

as the entropy of the partition. We then define the entropy of φ with respect to the partition as

$$h_\mu(\varphi, \mathcal{P}) = \lim_{n \to \infty} \frac{1}{n} H\left(\bigvee_{j=0}^{n} \varphi^{-j}(\mathcal{P})\right).$$

Recall that if $\mathcal{P}_0, \dots, \mathcal{P}_n$ are partitions, then $\bigvee_{j=0}^{n} \mathcal{P}_j$ is the partition whose elements are sets of the form $P_0 \cap \cdots \cap P_n$ for $P_j \in \mathcal{P}_j$ and $\mu(P_0 \cap \cdots \cap P_n) > 0$. *Finally, we define $h_\mu(\varphi)$ the (metric or measure theoretic) entropy of φ with respect to μ, as the supremum of $h_\mu(\varphi, \mathcal{P})$ over all finite partitions of* supp μ. Notice that $h_\mu(\varphi^{-1}) = h_\mu(\varphi)$.

This concept is also based on the intuitive notion of entropy that we have first presented. Instead of pursuing this point, which was quite clear with respect to the topological entropy, we now state the following important fact: *the topological entropy of φ is the supremum of the metric entropies of φ with respect to all invariant probability (ergodic) measures.* As a consequence, if $h(\varphi) > 0$ then there are invariant probability measures, even ergodic ones, with respect to which the metric entropy is also positive. We refer the reader to [**B**,1975a], [**M**,1987], [**S**,1976], [**W**,1982] for more references, details and some history.

We now present a theorem of Oseledec [**O**,1968], the multiplicative ergodic theorem, that together with subsequent work of Pesin ([**P**,1976], [**P**,1977]) has very much influenced a basic line of research that allows one to extend results valid for hyperbolic diffeomorphisms, e.g. the stable manifold theory, to other classes of maps.

We say that $x \in M$ is a regular point for φ if there are numbers $\lambda_1(x) > \cdots > \lambda_\ell(x)$ and a decomposition

$$T_x M = E_1(x) + \cdots \oplus E_\ell(x)$$

such that

$$\lim_{n \to \pm\infty} \frac{1}{n} \log \|(d\varphi^n)_x v\| = \lambda_j(x)$$

for $v \in E_j$, $v \neq 0$, and $1 \leq j \leq \ell$. The numbers $\lambda_j(x)$ are called Lyapunov exponents of φ at x. The multiplicative ergodic theorem states that for every invariant probability measure μ, the set of regular points Λ is a Borel subset and $\mu(\Lambda) = 1$. Moreover, the maps $x \to \lambda_j(x)$ and $x \to E_j(x)$ are measurable. If μ is ergodic, then the λ_j's as well as the dimensions of the corresponding E_j's are constant a.e. (almost everywhere).

We say that φ is *nonuniformly hyperbolic* if $\lambda_j(x) \neq 0$, x a.e. in Λ and all $1 \leq j \leq \ell$.

For nonuniformly hyperbolic C^2 (or even $C^{1+\alpha}$, $\alpha > 0$) diffeomorphisms, Pesin developed a "filtered" hyperbolic theory for the set Λ of regular points. That is, there are closed, but in general not invariant, hyperbolic sets $\Lambda_1 \subset \cdots \subset \Lambda_k \subset \cdots$, $k \in \mathbb{N}$ and $\bigcup_k \Lambda_k = \Lambda$. The main difference with the usual hyperbolic structure on Λ is that, as k increases, it may take longer and longer to detect the hyperbolic behaviour of $d\varphi$ on $E^u(x)$ and $E^s(x)$ for $x \in \Lambda_k$, where $E^u(x)$ and $E^s(x)$ denote the subspaces generated by the positive and negative Lyapunov exponents at x. Assuming that the measure is ergodic and denoting by $\lambda^+ = \min \lambda_j$, $\lambda_j > 0$, and by $\lambda^- = \max \lambda_j$, $\lambda_j < 0$, we have

$$\|(d\varphi^n)_x v\| \geq k^{-1} e^{\lambda^+ - \varepsilon} \|v\|$$
$$\|(d\varphi^n)_x w\| \leq k e^{\lambda^- + \varepsilon} \|w\|$$

for $x \in \Lambda_k$, $v \in E^u(x)$, $w \in E^s(x)$, $n \in \mathbb{N}$ and $\varepsilon > 0$ small. Moreover, the stable and unstable sets of $x \in \Lambda$ are injectively immersed submanifolds of M, denoted, as in the hyperbolic case, by $W^s(x)$ and $W^u(x)$. (Notice, however, that in this case the angles between $E^u(x)$ and $E^s(x)$, $x \in \Lambda_k$, may go to zero as $k \to \infty$.)

Thus, in the nonuniformly hyperbolic case, many vectors will grow exponentially when iterating φ. Vectors can be interpreted as "two infinitely close points" and, hence, we get exponential growth of their distances, which implies, informally speaking, exponential sensitive dependence on the initial conditions. It seems reasonable to even expect that full sensitivity is formally implied just by the existence of a positive Lyapunov exponent.

Based on the work of Pesin, Katok [**K**,1980] proved that the maximal invariant set in each Λ_k is hyperbolic with a dense subset of hyperbolic periodic points all of them exhibiting transversal homoclinic orbits. This pretty result corroborates our previous comments about the sensitivity of the orbits. Moreover, since in general we have

$$h_\mu \leq \sum_{\lambda_j > 0} \lambda_j,$$

known as Ruelle's inequality [**R**,1978], we conclude that if $\dim M = 2$ and $h(\varphi) > 0$ then φ is nonuniformly hyperbolic with respect to some ergodic probability measure μ. In fact, there is μ such that $h_\mu(\varphi)$ is close to $h(\varphi)$ and thus $h_\mu(\varphi) > 0$. This implies the existence of a strictly positive Lyapunov exponent by Ruelle's inequality. Similarly, since $h_\mu(\varphi^{-1}) = h_\mu(\varphi)$, we get a strictly negative Lyapunov exponent, proving the assertion. Notice that, for invariant ergodic measures that are absolutely continuous with respect to the Lebesgue measure, Pesin [**P**,1977] had proved that

$$h_\mu = \sum_{\lambda_j > 0} \lambda_j.$$

The previous theory is valid for any invariant measure, although often we considered ergodic measures to simplify some statements. There are, however, measures that describe the asymptotic behaviour of large classes of orbits, like the ones whose ω-limit set is an attractor. To describe such measures, let A be an attractor for the map φ with basin $B(A)$. For any continuous function g on M, one can consider its average along an orbit starting at $x_0 \in B(A)$

$$\lim_{n \to \infty} \frac{1}{n} \sum_{i=0}^{n-1} g(\varphi^i(x_0)).$$

Let us assume that the limit exists Lebesgue-a.e. in $B(A)$ and that it is independent of x_0 (such an assumption is commonly taken for granted in Physics). This averaging limit is a linear positive and normalized operator from the continuous functions $C^0(M)$ to $\mathbb{R}$ and, hence, defines a probability measure. The measure μ clearly has its support in A and it is a probability measure. Furthermore,

$$\lim_{n \to \infty} \frac{1}{n} \sum_{i=0}^{n-1} g(\varphi^i(x_0)) = \int g \, d\mu$$

for every continuous function g and a.a. $x_0 \in B(A)$.

This measure was introduced by Sinai and proved to exist for Anosov diffeomorphisms [**S**,1968], [**S**,1970]; the result was later extended to Axiom A systems by Ruelle and Bowen [**R**,1976], [**BR**,1975]. These measures are now called SBR measures. They were shown to exist for the expanding model of the Lorenz attractor by Bunimovich and Sinai [**BS**,1979], Afraimovich and Pesin [**AP**,1987] and for nonhyperbolic DA-type attractors in [**C**,1991].

The existence of SBR measures is a hard mathematical problem for many attractors. This fact did lead to a number of variations of the original definition. In particular, for certain Hénon-like attractors of the family of quadratic diffeomorphisms, Benedicks and Young [**BY**,1991] exhibit an ergodic measure supported on the attractor and with respect to which the map is nonuniformly hyperbolic. Furthermore, the disintegration of this measure along the unstable manifolds is absolutely continuous with respect to the Lebesgue measure. This is indeed a reasonable variation of the original definition since the construction of the SBR measure in the Axiom A case was done through integrating absolutely continuous measures along the unstable manifolds in the attractor. Also, the limit property concerning the average of continuous functions along orbits holds at least for Lebesgue almost all points in some open set.

We end this section by posing the following problem.

PROBLEM 2: Let φ be a C^2 diffeomorphism and μ an ergodic probability measure, invariant under φ, such that if supp $\mu = A$, then A is a nonuniformly hyperbolic attractor and μ desintegrates into absolutely continuous measures along the unstable manifolds. Is it true that the union of the stable manifolds of μ-almost all points in A covers the basin $B(A)$ up to a set of Lebesgue measure zero? In particular, is this the case for the Hénon-like attractors discussed before?

§3   Summary, further results and problems

Let us summarize the results discussed in previous sections and chapters on unfoldings of homoclinic tangencies. We shall add comments and pose questions that might, in the context of our programme, be of relevance to achieving new levels of understanding of such a phenomenon.

As before, let φ_μ be a family of locally dissipative diffeomorphisms unfolding generically a quadratic homoclinic tangency of $\varphi_0 = \varphi_{\mu=0}$. Suppose that this tangency is associated with a periodic point p_0 and a basic set $\Lambda_0 \ni p_0$. As in Section 2, Chapter 5, let $U_\mu = U_1 \cup U_\mu^*$, where

- U_1 is a small fixed neighbourhood of Λ_0,

- $U_\mu^* = \bigcup_{|i| \le N} \varphi_\mu^i(V_\mu)$, V_μ is a neighbourhood of a point in the orbit of tangency whose points are at a distance at most $K\mu$ from both the local stable and unstable manifolds of p_μ, the continuation of p_0, for some constant $K > 0$,

so that U_μ is a neighbourhood of both Λ_0 and the orbit of tangency. Also, define the thickness of Λ_0 by $\tau(\Lambda_0) = \tau^s(\Lambda_0) \cdot \tau^u(\Lambda_0)$, where $\tau^s(\Lambda_0)$ is the stable thickness and $\tau^u(\Lambda_0)$ the unstable thickness. Denote by Λ_μ the (hyperbolic) continuation of Λ_0 and by $B(\{\varphi_\mu\})$ or simply by $B(\varphi_\mu)$ the set of parameter values for which the *maximal invariant set* of $\varphi_\mu \,|\, U_\mu$ is not hyperbolic. We then have, for $\varepsilon > 0$ small and $-\varepsilon < \mu < \varepsilon$, the following

- There are intervals I_i in $(-\varepsilon, \varepsilon)$ and residual subsets $N_i \subset I$, $i \in \mathbf{N}$, such that φ_μ has infinitely many sinks for each $\mu \in N_i$. Moreover, if $\tau(\Lambda_0) > 1$, then some stable and unstable manifolds of points in Λ_μ, for every $\mu \in (-\varepsilon, 0)$ or $\mu \in (0, \varepsilon)$, are tangent to each other (Newhouse),

- There is a positive Lebesgue measure set $E \subset (-\varepsilon, \varepsilon)$ such that φ_μ has a Hénon-like strange attractor for each $\mu \in E$ (Mora and Viana, based on Benedicks and Carleson),

– If $HD(\Lambda_0) < 1$, then

$$\lim_{\varepsilon \to 0} \frac{m(B(\varphi_\mu) \cap (-\varepsilon, \varepsilon))}{\varepsilon} = 0.$$

– If $HD(\Lambda_0) > 1$, then for most families $\{\varphi_\mu\}$ a partial converse is true, namely

$$\limsup_{\varepsilon \to 0} \frac{m(B(\varphi_\mu) \cap (-\varepsilon, \varepsilon))}{\varepsilon} > 0.$$

So, in terms of dynamical prevalence, we have something like Figure 7.5. Notice that part of the figure is just symbolic, since we cannot translate the condition $\tau^s(\Lambda_0) \cdot \tau^u(\Lambda_0) > 1$ into $HD(\Lambda_0) > a > 1$.

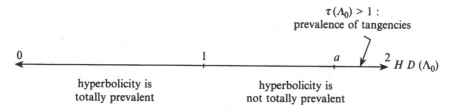

Figure 7.5

The last two of the above results, proved in [**PT**,1987] and [**PY**,1991], respectively, bring up the importance of the Hausdorff dimension of the basic set associated with the homoclinic tangency for "measuring" the bifurcation set of $\{\varphi_\mu\}$ near φ_0. *Indeed, if $HD(\Lambda_0) < 1$, then hyperbolicity is the prevalent dynamical phenomenon near φ_0 along "typical" curves of diffeomorphisms near φ_0. We also say in this case that $\mu = 0$ is a point of (full) density of hyperbolicity for the family φ_μ. On the other hand, this is not the case for almost all one-parameter families of diffeomorphisms when $HD(\Lambda_0) > 1$.* It is relevant to know in this last case, i.e., $HD(\Lambda_0) > 1$, if typically $\mu = 0$ is or not a point of positive density of hyperbolicity, or of Hénon-like strange attractors, or of infinitely many coexisting sinks.

PROBLEM 1: Is it true, for a family φ_μ as above, that $\mu = 0$ is a point of density zero for the set of μ's for which φ_μ has a Hénon-like strange attractor when $HD(\Lambda_0) > 1$?

PROBLEM 2: The same as in the previous problem, but now for periodic attractors (sinks). How about infinitely many sinks? These questions are particularly interesting when $\tau(\Lambda_0) > 1$.

As we saw, residually in some intervals in the μ-parameter line, there are infinitely many sinks for the corresponding maps φ_μ. *For the Hénon-like strange attractors we can ask if it is possible that infinitely many of them occur simultaneously when a homoclinic tangency is unfolded.* In any case, we have the following conjecture stating that the simultaneous existence of infinitely many sinks or Hénon-like strange attractors might be a "rare" phenomenon.

CONJECTURE 1. *The set of μ's for which φ_μ has infinitely many sinks or infinitely many Hénon-like strange attractors has Lebesgue measure zero.*

Before this conjecture was formulated, Tedeschini-Lalli and Yorke [**TY**, 1986] proved that the set of parameters corresponding to infinitely many "simple" sinks has Lebesgue measure zero. Although their result dealt only with the simplest kind of sinks that are created when unfolding homoclinic tangencies, the general question on the Lebesgue measure of the set of parameter values corresponding to infinitely many sinks was implied there.

Concerning the condition on the thickness of Λ_0, $\tau(\Lambda_0) > 1$, to ensure the existence of tangencies between leaves of $\mathcal{F}^s(\Lambda_\mu)$ and $\mathcal{F}^u(\Lambda_\mu)$ for every $\mu \in (-\varepsilon, 0)$ or $\mu \in (0, \varepsilon)$, one can ask if the condition is necessary to achieve this persistence of tangencies. In work in development, Yoccoz and the first author of this book show that this is not the case for an open set of C^2 diffeomorphisms. This leads to the following very interesting question.

PROBLEM 3: Is there a necessary and sufficient condition, say involving fractal dimensions of the basic set Λ_0, to obtain an interval of the form $(-\varepsilon, 0)$ or $(0, \varepsilon)$ such that for each $\mu \in (-\varepsilon, 0)$ or $\mu \in (0, \varepsilon)$ there is a tangency between leaves of the stable and unstable foliations of Λ_0?

Related to Problem 1 there is a relevant recent result of Diaz, Rocha and Viana [**DRV**,1991], stating that a *saddle-node critical cycle* corresponds to a point of *positive density* of Hénon-like strange attractors for generic one-parameter families of surface diffeomorphisms. That is, if $SA(\varphi_\mu)$ indicates the set of μ's for which φ_μ exhibits Hénon-like strange attractors and φ_0 has a saddle-node critical cycle, then

$$\liminf_{\varepsilon \to 0} \frac{m(SA(\varphi_\mu) \cap (-\varepsilon, \varepsilon))}{\varepsilon} > 0.$$

It is interesting to note that, by a remark of Mora, saddle-node critical cycles are always present when generically unfolding a homoclinic tangency. This can be seen as follows. As first pointed out in the last section of Chapter 3 and much used in Chapter 6, we can study properties of φ_μ that are "robust" under perturbations of the interval map $y \to 1 - ay^2$, since for some fixed integer N this is the limiting map as $n \to \infty$ of φ_μ^{n+N} restricted to suitable domains. We then notice that for $a = 1.75$ the interval map has a saddle-node whose unstable separatrix contains a critical point and also a pre-orbit of the saddle-node. Thus, for μ small, φ_μ has a saddle-node whose unstable separatrix intersects transversely its strong stable manifold. This implies that, for μ small, there is the collapse of a sink and a saddle which is part of a horseshoe, like in Figure 7.1(b).

Let us consider the case of a saddle-node critical 1-cycle, like in Figure 7.1(a). Let C be a small φ_μ-positively invariant annulus containing in its interior the saddle-node p_0 and its unstable separatrix. Notice that the unstable separatrix (including p_0) forms a topological circle.

PROBLEM 4: Is $\mu = 0$ a point of positive density of hyperbolicity for $\varphi_\mu \,|\, C$?

PROBLEM 5:[3] Are there values of μ near zero for which $\varphi_\mu \,|\, C$ has a unique *global attractor* which is the closure of some unstable separatrix and which is persistent ?

PROBLEM 6: Find other dynamical configurations, besides the saddle-nodes critical cycles, that correspond to a point of positive density of Hénon-like strange attractors.

Motivated by the unfolding of homoclinic tangencies associated with basic sets, we were led to the following questions on the arithmetic difference of dynamically defined or regular Cantor sets on the line.

PROBLEM 7: Is it true that if C_1 and C_2 are affine Cantor sets, then either $C_1 - C_2$ has Lebesgue measure zero or else it contains intervals ? We *conjecture* that this should be the case for "most" affine Cantor sets C_1 and C_2. Maybe this is true even for all pairs of affine Cantor sets!

PROBLEM 8: The same as the previous problem but now for "most" dynamically defined Cantor sets.

In this direction Sannami [S,1991] has given an example of dynamically defined Cantor sets whose arithmetic difference has positive measure but contains no intervals. Also, very interesting is the work of Larsson [L,1990] who showed that the above conjecture is true for what he calls "random" Cantor sets which, however, are not regular in our sense. On the other hand, Mendes and Oliveira have work in development about the topological structure of $C_1 - C_2$, C_1 and C_2 being affine.

We now pose some questions motivated by Conjecture 1 in Section 1 of this chapter.

PROBLEM 9: Suppose that $\{\varphi_n\}$ is a sequence of C^k surface diffeomorphisms C^k-converging to φ, $k \geq 3$, such that each φ_n exhibits a period doubling bifurcation yielding sinks of period $p \cdot 2^n$, for some $p \in \mathbf{N}$. Can φ be C^k-approximated by a diffeomorphisms having a homoclinic tangency ?

---

[3]Added in proof: Diaz, Rocha and Viana have announced that this is so for some open set of one-parameter families of diffeomorphisms.

Similarly, if φ exhibits infinitely many sinks or if φ has a Hénon-like strange attractor. As noticed above in Section 2 the answer is positive for the "known" Hénon-like attractors by the work of Ures [**U**,1992]. Also, we do not add the critical saddle-node cycle to this list because the answer is known to be true [**NPT**,1983]. The same questions can be posed in higher dimensions.

COMMENT: The phenomena mentioned in this problem are probably the most important that have been discovered so far for surface diffeomorphisms. A positive answer to the questions would certainly give much strength to our Conjecture 2 in Section 1 concerning the density among the nonhyperbolic surface diffeomorphisms of the ones exhibiting a homoclinic tangency.

PROBLEM 10: Suppose that $\{\varphi_\mu\}$, $\mu \in \mathbb{R}$, is a generic family of surface diffeomorphisms such that $h(\varphi_\mu) = 0$ for $\mu < 0$ and $h(\varphi_\mu) > 0$ for $\mu \geq 1$, where $h(\varphi_\mu)$ indicates the topological entropy of φ_μ. Show that there is $0 \leq \mu_0 \leq 1$ such that φ_{μ_0} has a homoclinic tangency. Similarly, if the limit set $L(\varphi_\mu)$ consists for $\mu < 0$ of finitely many periodic orbits and for $\mu > 1$ exhibits an invariant horseshoe.

Let us comment briefly that the first three results recalled in this section for families of surface diffeomorphisms, concerning *the coexistence of infinitely many sinks, the existence of Hénon-like strange attractor and the relative prevalence of hyperbolicity in the presence of a basic set of Hausdorff dimension smaller than 1*, are now also available in *higher dimensions* (Palis and Viana [**PV**,1991], Viana [**V**,1991b] and Takens [**T**,1991a], [**T**,1991b]), respectively. It is to be noted that the first two were proved for unfoldings of codimension-1 homoclinic tangencies, i.e. when the dimension of the unstable manifold of the associated saddle is 1. To obtain attractors, sinks and Hénon-like attractors, one assumes that the product of any two eigenvalues at the associated saddle to have norm less than 1. We believe that the fourth result mentioned in the same list is also true in the codimension-1 case: hyperbolicity is not of full density at a homoclinic tangency in the presence of a basic set of Hausdorff dimension bigger than 1.

The same proof as in Chapter 3 also yields cascades of period doubling bifurcations of sinks in the codimension-1 case if the product of any pair of eigenvalues has norm less than 1. (See Franks [**F**,1985] for a related result). In the general case of unfolding a homoclinic tangency of any codimension, the corresponding result showing the existence of cascades of period doubling bifurcations of saddles has recently been obtained by Martín–Rivas [**M**,1991]. As for the persistence of tangencies in the general case, Romero [**R**,1992], based on the codimension-1 case in [**PV**,1991], has recently given an affirmative answer: there is an open set of nearby diffeomorphisms and a dense subset of it whose elements exhibit a homoclinic tangency. (See

Chapter 6).

A source of difficulties in higher dimensions is the fact that in general the stable (unstable) foliation of a basic set is no longer differentiable unless it is of codimension-1. *Actually, a good question is whether the local definition of Hausdorff dimension for a basic set in higher dimension is independent of the point.*

In line with our programme to study non-hyperbolic dynamics, we close this chapter with a general question concerning generic (most, residual) one-parameter families of surface diffeomorphisms: *Is it true that the maps exhibit sinks or quasi-periodic attractors or else strange attractors for Lebesgue almost all parameter values? The same for endomorphisms of the interval (circle) or rational mappings of the Riemann sphere concerning sinks or absolutely continuous invariant measures.*

APPENDIX 1

HYPERBOLICITY: STABLE MANIFOLDS AND FOLIATIONS

The purpose of this appendix is to collect a number of results from the literature, especially from [**HPS**,1977], which were used in the previous chapters. These theorems are all related to the notion of hyperbolicity (norms of eigenvalues being different from 1). They are based on the construction of objects which are invariant under a diffeomorphism φ, by applying φ^n to a noninvariant object and then taking a limit. We present sketches and main ideas of proofs that might help the reader to understand these basic results (and perhaps even guess and prove other results in the same direction).

STABLE MANIFOLD THEOREM 1. *Let $\varphi\colon M \to M$ be a C^k diffeomorphism, $k \geq 1$, and let $p \in M$ be a fixed point, i.e. $\varphi(p) = p$, such that $(d\varphi)_p$ has no eigenvalue of norm 1 (in this case one calls p a hyperbolic fixed point). Then the stable manifold*

$$W^s(p) = \{x \in M \mid \lim_{n\to+\infty} \varphi^n(x) = p\}$$

is an injectively immersed C^k submanifold of M. If $\tilde{\varphi}$ is C^k near φ, $\tilde{\varphi}$ has a hyperbolic fixed point $\tilde{p}$ near p and the stable manifold $W^s(\tilde{p})$ is near $W^s(p)$ in the C^k sense, at least if we restrict ourselves to compact neighbourhoods of $\tilde{p}$ and p in $W^s(\tilde{p})$ and $W^s(p)$.

Applying the same theorem to φ^{-1}, we obtain the *unstable manifold* $W^u(p)$. One way to prove the theorem is to find $W^s(p)$ as the fixed point of the so called *graph transform*. We first describe the local construction; at this point we may assume that M is a euclidean space and $p = 0$. Let $T_pM = E^s \oplus E^u$, where E^s and E^u denote the eigenspaces of $(d\varphi_p)$, corresponding to the eigenvalues whose norm is less than 1 and greater than 1, respectively. Let B^s and B^u be small neighbourhoods of $p = 0$ in E^s and E^u. Denote by $C(B^s, B^u)$ the space of continuous maps (to be formally correct: Lipschitz maps with Lipschitz constant ≤ 1) $u\colon B^s \to B^u$, endowed with the C^0 norm defined by $\|u\| = \sup\{u(x) \mid x \in B^s\}$. If $W = $ graph (u) is the graph of such a map, then the same holds for $\varphi^{-1}(W)$ restricted to $B^s \times B^u$: there is $u_1\colon B^s \to B^u$ such that graph $(u_1) = \varphi^{-1}(W) \cap (B^s \times B^u)$. It is not difficult to see that the graph transform $\Gamma_\varphi\colon u \mapsto u_1$, defined in this way, is a contraction in $C(B^s, B^u)$. It follows that there is a unique continuous map $u_\varphi\colon B^s \to B^u$ such that $\Gamma_\varphi(u_\varphi) = u_\varphi$ and $u_\varphi = \lim_{n\to+\infty} \Gamma_\varphi^n(u)$ for every $u \in C(B^s, B^u)$. The differentiability of u_φ is a more delicate question. Although the subspace H of C^k maps $u: B^s \to B^u$ satisfies $\Gamma_\varphi(H) \subset H$, it does not follow that $u_\varphi \in H$, because H is not closed in $C(B^s, B^u)$. On the other hand, in general Γ_φ is not a contraction relative to C^k-norms of C^k maps ($\|u\|_k = \sup\{\|d^i u(x)\| \, ; \, x \in B^s, \, 0 \leq i \leq k\}$). Nevertheless, one can prove that u_φ is a C^k map, but this requires a more sophisticated argument. We will return to this point in a more general context when discussing normally hyperbolic submanifolds.

Let now W_φ denote the graph of u_φ. By construction, this is a C^k embedded disk containing p, with $T_p W_\varphi = E^s$, and it is φ-invariant in the sense that $\varphi(W_\varphi) \subset W_\varphi$. Moreover it is not difficult to check that W_φ contains precisely the points of $B^s \times B^u$ whose positive iterates remain forever in $B^s \times B^u$ and that these iterates, in fact, converge to p. (In other words: W_φ is a *local stable manifold* of p.) Then we must have $W^s(p) = \bigcup_{n \geq 0} \varphi^{-n}(W_\varphi)$, which proves that $W^s(p)$ is an injectively immersed C^k submanifold of M as stated.

Finally, nearby diffeomorphisms induce nearby graph transforms, which then have nearby fixed points. Thus, the stable manifolds of these diffeomorphisms are close to each other on compact parts in the C^0 sense. Moreover, from the proof that the stable manifold is C^k, one concludes that this proximity can actually be taken in the C^k sense, which finishes our sketch of proof of the stable manifold theorem.

It is also worth observing that the argument in the proof is true in general Banach spaces. (*Compact parts* in the statement is replaced by *bounded parts*). Hence, it proves the theorem even when M is infinite-dimensional.

Our next result, the so called λ-lemma or inclination lemma, relies on the same basic idea as the previous theorem; constructing the stable manifold as the limit of backward iterates $\varphi^{-n}(W)$ of submanifolds of the right dimension. Again, it holds in general Banach manifolds.

THEOREM 2 [**P**,1969]. *Let $\varphi: M \to M$ be a C^k diffeomorphism, $k \geq 1$, with a hyperbolic fixed point p. Let $W \subset M$ be a C^k submanifold such that*

- *$\dim(W) = \dim(W^s(p))$,*

- *W has a point q of transversal intersection with $W^u(p)$.*

Then $\varphi^{-n}(W)$ converges to $W^s(p)$ in the following sense. For each n one can choose a disk $D_n \subset \varphi^{-n}(W)$, which is a neighbourhood of $\varphi^{-n}(q)$ in $\varphi^{-n}(W)$, such that

$$\lim_{n \to \infty} D_n = D$$

where D is a disk-neighbourhood of p in $W^s(p)$. Here, the convergence means that for n big enough D_n and D are C^k-near embedded disks.

The λ-lemma can be used to give a geometric construction of stable and unstable foliations in a neighbourhood of a hyperbolic saddle point p. By this we mean invariant foliations $\mathcal{F}^s$ and $\mathcal{F}^u$ which are transversal to each other near p and such that the leaves through p satisfy $\mathcal{F}^s(p) \subset W^s(p)$, $\mathcal{F}^u(p) \subset W^u(p)$. We only describe the construction of $\mathcal{F}^s$; $\mathcal{F}^u$ is obtained by applying the same procedure to φ^{-1} instead of φ.

First we take $N \subset W^u(p)$ such that every point of $W^u(p)$ has at least one iterate in the interior of N, and we C^k-foliate a neighbourhood of N by disks transversal to $W^u(p)$. In other words, we take D a disk with $\dim(D) = \dim(W^s(p))$ and we consider a C^k diffeomorphism $\Phi: N \times D \to M$ such that for every $x \in N$, $\Phi(x, 0) = x$ and $\mathcal{F}^s(x) = \Phi(x, D)$ is an embedded disk transversal to $W^u(p)$. This can be done in such a way that $\varphi^{-1}(\mathcal{F}^s(x)) \supset \mathcal{F}^s(\varphi^{-1}(x))$ whenever x and $\varphi^{-1}(x)$ both belong to N. Then, we can extend $\mathcal{F}^s$ to an invariant foliation near p (which we still denote $\mathcal{F}^s$) by defining, for $n \geq 0$ and $x \in N$, $\mathcal{F}^s(\varphi^{-(n+1)}(x)) = \varphi^{-1}(\mathcal{F}^s(\varphi^{-n}(x))) \cap V$ and $\mathcal{F}^s(p) = W^s(p) \cap V$, for some fixed neighbourhood V of p.

The λ-lemma implies that in this way we obtain a *continuous* foliation in a neighbourhood of p. We want to make a remark concerning its differentiability. Clearly all the leaves of $\mathcal{F}^s$ are C^k submanifolds. Moreover, it follows immediately from our construction that the projections along leaves of $\mathcal{F}^s$ are C^k at every point outside $\mathcal{F}^s(p)$. But this is no longer true for the points in $\mathcal{F}^s(p)$: the projection along the leaves of $\mathcal{F}^s$ is in general *not differentiable* at these points. So in general $\mathcal{F}^s$ is not even a C^1 foliation. However, later we will show that, if $\mathcal{F}^s$ has codimension 1 (e.g. if M is a surface) and if φ is C^2 then $\mathcal{F}^s$ is indeed a C^1 foliation. If moreover φ is C^3 then $\mathcal{F}^s$ is even $C^{1+\varepsilon}$ for some $\varepsilon > 0$.

The key hypothesis in the results above is the hyperbolicity of the fixed point p, i.e. the existence of a $d\varphi$-invariant splitting $T_p M = E^s \oplus E^u$ such that $d\varphi_p \mid E^s$ is contracting and $d\varphi_p \mid E^u$ is expanding. Now we want to extend this notion to invariant subsets of M which do not necessarily consist of a single point, in order to derive similar properties for such sets. Let $\varphi: M \to M$ be a C^k diffeomorphism, $k \geq 1$. A compact set $\Lambda \subset M$ is a *hyperbolic set for* φ if $\varphi(\Lambda) = \Lambda$ and if there is a splitting $T_x(M) = E_x^u \oplus E_x^s$ for each $x \in \Lambda$ such that

– the splitting depends continuously on x,

– the splitting is invariant, i.e. $d\varphi(E_x^u) = E_{\varphi(x)}^u$ and $d\varphi(E_x^s) = E_{\varphi(x)}^s$,

– there are constants $\sigma > 1$ and $C > 0$ such that for any $v \in E_x^u$ and $n > 0$,

$$\|d\varphi^n(v)\| \geq C \cdot \sigma^n \cdot \|v\|,$$

and such that for any $v \in E_x^s$ and $n > 0$,

$$\|d\varphi^n(v)\| \leq C^{-1} \cdot \sigma^{-n} \cdot \|v\|.$$

Here $\| \quad \|$ denotes the norm of tangent vectors of M with respect to some fixed Riemannian metric on M.

Note that a hyperbolic fixed point is a special case of a hyperbolic set. On the other hand, in a deeper sense, a hyperbolic set always corresponds to a hyperbolic fixed point of a mapping in some infinite-dimensional manifold. Let us explain this affirmative. Denote by $C(\Lambda, M)$ the space of continuous mappings $f \colon \Lambda \to M$, where Λ is some compact φ-invariant part of M. This space can be endowed with a natural structure of differentiable manifolds modelled on the Banach space of vector fields of M over Λ. Relative to this structure, the operator $F_\varphi \colon C(\Lambda, M) \hookleftarrow$, $F_\varphi(f) = \varphi f \varphi^{-1}$ is of class C^k, if φ is C^k. Clearly, the inclusion $i \colon \Lambda \hookrightarrow M$ is a fixed point for F_φ. It turns out that this fixed point is hyperbolic if and only if Λ is a hyperbolic set for φ.

This construction may be used to derive properties of hyperbolic sets from corresponding facts for hyperbolic fixed points. A good example of this is the following result concerning persistence of hyperbolic sets under small perturbations.

THEOREM 3. *Let $\Lambda \subset M$ be a hyperbolic set for a C^k diffeomorphism φ, $k \geq 1$. Then, for every diffeomorphism Ψ which is C^k close to φ, there is a unique homeomorphism $h_\Psi \colon \Lambda \to M$ C^0-close to the inclusion $i \colon \Lambda \hookrightarrow M$, such that $\Lambda_\Psi = h_\Psi(\Lambda)$ is a hyperbolic set for Ψ and $\Psi \circ h_\Psi = h_\Psi \circ \varphi$ on Λ. This conjugacy h_Ψ is Hölder continuous. Moreover, if Λ is the maximal φ-invariant set in a neighbourhood U of Λ, then Λ_Ψ is the maximal Ψ-invariant set in U.*

To prove this, we observe that $G_\Psi \colon C(\Lambda, M) \hookleftarrow$ defined by $G_\Psi(f) = \Psi f \varphi^{-1}$ is close to $F_\varphi(f) = \varphi f \varphi^{-1}$ and so it must have a hyperbolic fixed point, say h_Ψ, near i. In fact, i is a hyperbolic fixed point for F_φ and such points persist under small perturbations. The injectivity of h_Ψ is a consequence of the fact that $\varphi \mid \Lambda$ is expansive, i.e. there is $\varepsilon > 0$ such that for $x, y \in \Lambda$, $x \neq y$, we have $\rho(\varphi^n(x), \varphi^n(y)) > \varepsilon$ for some $n \in \mathbf{Z}$, where ρ denotes the distance with respect to some Riemann metric on M. In fact $h_\Psi(x) = h_\Psi(y)$ implies $h_\Psi(\varphi^n(x)) = h_\Psi(\varphi^n(y))$ for all $n \in \mathbf{Z}$. Then if Ψ is close enough to φ (so that $\rho(h_\Psi(x), x) < \varepsilon/2$ for all $x \in \Lambda$), it follows that $\rho(\varphi^n(x), \varphi^n(y)) \leq \varepsilon$ for all $n \in \mathbf{Z}$ and so $x = y$. Finally, let $F_\Psi \colon C(\Lambda_\Psi, M) \hookleftarrow$, $F_\Psi(g) = \Psi \circ f \circ \Psi^{-1}$, and $H_\Psi \colon C(\Lambda_\Psi, M) \to C(\Lambda, M)$, $H_\Psi(g) = g \circ h_\Psi$. Then $F_\Psi = H_\Psi^{-1} \circ G_\Psi \circ H_\Psi$ and so $dF_\Psi(i_\Psi) = (dH_\Psi(i_\Psi))^{-1} \circ dG_\Psi(h_\Psi) \circ (dH_\Psi(i_\Psi))$, where $i_\Psi \colon \Lambda_\Psi \hookrightarrow M$ is the inclusion. Since h_Ψ is a hyperbolic fixed point for G_Ψ, the same holds for i_Ψ and F_Ψ, which proves that Λ_Ψ is a hyperbolic set for Ψ. Finally, if $\Lambda = \bigcap_{n \in \mathbf{Z}} \varphi^n(U)$ for some neighbourhood U of Λ, then, clearly, $\Lambda_\Psi = h_\Psi(\Lambda) \subset \bigcap_{n \in \mathbf{Z}} \Psi^n(U)$. One can prove the reverse inclusion using the technique of shadowing of orbits [S,1978], [B,1977]. This proves the theorem.

Another application of this idea is the following generalization of the stable manifold theorem.

GENERALIZED STABLE MANIFOLD THEOREM 4. *Let $\Lambda \subset M$ be a hyperbolic set for a C^k diffeomorphism φ, $k \geq 1$. Then for each $x \in \Lambda$ the stable manifold*

$$W^s(x) = \{y \in M \mid \lim_{n \to +\infty} \rho(\varphi^n(x), \varphi^n(y)) = 0\}$$

is an injectively immersed C^k submanifold. Moreover, $W^s(x)$ depends continuously on x in the sense that there is a continuous map $\Phi \colon \Lambda \to$ $\mathrm{Emb}^k(D, M)$ such that for each $x \in \Lambda$, the image of $\Phi(x)$ is a neighbourhood of x in $W^s(x)$. Here, $\mathrm{Emb}^k(D, M)$ is the space of C^k embeddings in M of a disk D whose dimension equals the dimension of E_x^s (which we assume to be independent of x). Finally, these stable manifolds also depend continuously on the diffeomorphism φ, in the sense that nearby diffeomorphisms yield nearby mappings Φ as above.

The proof of this result can be reduced to the case of a fixed point through the relation $W^s(x) = ev_x(W^s(i)) = \{f(x) \mid f \in W^s(i)\}$, where $W^s(i)$ denotes the stable manifold of the inclusion $i \colon \Lambda \hookrightarrow M$ as a fixed point of the operator F_φ defined above.

In the above situation (hyperbolic sets and their collection of stable manifolds) it is often useful to use *local stable manifolds* which are defined as follows

$$W_\varepsilon^s(x) =$$
$$\{y \in M \mid \lim_{n \to \infty} \rho(\varphi^n(x), \varphi^n(y)) = 0 \text{ and } \rho(\varphi^n(x), \varphi^n(y)) \leq \varepsilon \text{ for all } n \geq 0\}.$$

These local stable manifolds are, for ε sufficiently small, embedded disks which depend continuously on x in the C^k sense. Local unstable manifolds are defined analogously with φ replaced by φ^{-1}.

We come now to a different generalization of the stable manifold theorem. It provides many more invariant manifolds for fixed points (not necessarily hyperbolic) of diffeomorphisms.

THEOREM 5. *Let $\varphi \colon M \to M$ be a C^k diffeomorphism, $k \geq 1$ and $p \in M$, with $\varphi(p) = p$. Let $T_pM = E_1 \oplus E_2$ be a splitting such that for some $a > 0$,*

- $d\varphi(E_i) = E_i, \quad i = 1, 2,$
- *the norms of the eigenvalues of $d\varphi \mid E_1$ are smaller than a,*
- *the norms of the eigenvalues of $d\varphi \mid E_2$ are greater than a.*

Then there is a locally invariant C^1 manifold V with $p \in V$ and $T_p(V) = E_1$. It is locally invariant in the sense that $V \cap \varphi(V)$ contains a neighbourhood of p in V. The manifold V is in general not unique and not C^k for $k > 1$.

When we come to normally hyperbolic invariant manifolds, we shall say more about the *exact amount of differentiability* of (locally) invariant manifolds like V in the above theorem. At this point, however, we first want to indicate how these manifolds can also be obtained by a limiting process.

If one tries to reproduce the local argument in the proof of the stable manifold theorem (with E_1, E_2 in the place of E^s and E^u respectively), it may not work in this case. In fact, graphs of (local) functions are not necessarily transformed by φ^{-1} into graphs of functions having *at least the same domain*. Figure A1.1 illustrates this possibility for the construction of a two-dimensional invariant manifold containing a contracting and a weakly expanding direction of a three-dimensional saddle. So when constructing V we must consider two different situations. Assume first that all the eigenvalues of $d\varphi_p \mid E_1$ are contracting. Then we may, and do, apply the local construction used above for the stable manifold. In this case V is unique and depends continuously on the diffeomorphism. Suppose now that $d\varphi_p \mid E_1$ has also noncontracting eigenvalues. In this case one proceeds as follows. First, since the conclusions of the theorem are local we may assume that $M = \mathbb{R}^n$ and $p = 0$. Then we write $\varphi(x) = L(x) + \tilde{\varphi}(x)$, where $L = (d\varphi)_0$ and $\tilde{\varphi}(0) = 0$ and $(d\tilde{\varphi})_0 = 0$. Next, we choose a C^∞ function Ψ on $\mathbb{R}^n$ which is identically equal to 1 in a neighbourhood of the origin and is equal to zero outside the unit ball. We define $\overline{\varphi}(x) = L(x) + \Psi(x/\varepsilon) \cdot \tilde{\varphi}(x)$, where ε is positive and small enough so that $\overline{\varphi}$ is a diffeomorphism. Note that $\overline{\varphi}$ is equal to φ in a neighbourhood of the origin and that $\overline{\varphi} - L$ has small Lipschitz constant. Now one shows as before that the graph-transform of mappings $u: E_1 \to E_2$ is a contraction (this would not be true if $d\varphi_p \mid E_2$ had nonexpanding eigenvalues). Its fixed point u_φ corresponds to a submanifold $V = \text{graph}(u_\varphi)$ which is $\overline{\varphi}$-invariant and so locally invariant by φ, as required. This finishes the construction of V in this second case. Note that we may think of V as the limit of $\overline{\varphi}^{-n}(E_1)$ when $n \to +\infty$. It is clear that V depends on the several choices made in the construction; different choices may give rise to different locally invariant submanifolds V satisfying the conclusions in the theorem. On the other hand, for fixed choices the construction above yields nearby submanifolds when applied to nearby C^k diffeomorphisms. In this sense, V depends continuously on φ even in this case.

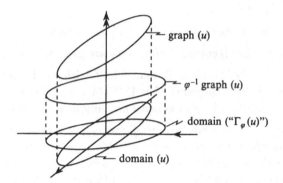

graph (u)

φ^{-1} graph (u)

domain ("$\Gamma_\varphi(u)$")

domain (u)

Figure A1.1

As a first application of the above theorem we construct now the *centre manifold*, the *centre-stable manifold* and the *centre-unstable manifold*. We begin with the centre-stable manifold. Let $\varphi\colon M \to M$ and $p \in M$ be as in the theorem. Take $a > 1$ such that an eigenvalue of $(d\varphi)_p$ has norm less than or equal to 1 if and only if it has norm less than a. For this $a > 1$ we construct the corresponding splitting $T_pM = E_1 \oplus E_2$ and apply the theorem. The resulting locally invariant manifold is called the *centre-stable manifold* and is denoted by $W^{cs}(p)$. As we will see, if φ is C^k, $1 \leq k < \infty$, then one can choose $W^{cs}(p)$ to be C^k. The *centre-unstable manifold* for φ is just the centre-stable manifold for φ^{-1}; it is denoted by $W^{cu}(p)$. The *centre manifold* is the intersection of a centre-stable and a centre-unstable manifold: $W^c(p) = W^{cs}(p) \cap W^{cu}(p)$.

Now we define the *strong stable manifold* and the *strong unstable manifold*. Let $0 < a < 1$ be such that $(d\varphi)_p$ has some eigenvalue with norm less than a and some other with norm in $(a, 1)$. Let $T_pM = E_1 \oplus E_2$ be the corresponding splitting. Applying the theorem we obtain an invariant manifold which we denote by $W^{ss}(p)$ and call the strong stable manifold. Unlike $W^c(p)$, $W^{cs}(p)$ and $W^{cu}(p)$, $W^{ss}(p)$ is unique (for a given $a < 1$) and it is characterized by being the set of points whose distance from p decreases exponentially at a rate stronger than $a < 1$. The strong stable manifold is C^k if φ is C^k, $k \geq 1$. The strong unstable manifold for φ is just the strong stable manifold for φ^{-1}.

The (locally) invariant manifolds which we have seen so far are all examples of *normally hyperbolic invariant manifolds*. This notion is defined as follows. Let $\varphi\colon M \to M$ be a C^k diffeomorphism, $k \geq 1$. A submanifold $V \subset M$ is said to be a normally hyperbolic invariant submanifold if $\varphi(V) = V$ and if there is a splitting $T_x(M) = T_x(V) \oplus N^s_x \oplus N^u_x$ for each $x \in V$ such that

– the splitting depends continuously on x,

– the splitting is invariant under $d\varphi$, i.e. $d\varphi(N_x^s) = N_{\varphi(x)}^s$ and $d\varphi(N_x^u) = N_{\varphi(x)}^u$,

– for some Riemannian metric, and constants $C > 0$, $\sigma > 1$, $r \geq 1$, one has for every triple of unit vectors $v \in T_x(V)$, $n^s \in N_x^s$, and $n^u \in N_x^u$ and any $n > 0$

$$\frac{\|(d\varphi^n)n^u\|}{\|(d\varphi^n)v\|^r} \geq C \cdot \sigma^n \text{ and } \frac{\|(d\varphi^n)n^s\|}{\|(d\varphi^n)v\|^r} \leq C^{-1} \cdot \sigma^{-n}.$$

In this case one calls V r-*normally hyperbolic*.

One of the reasons why the notion of hyperbolicity is of importance is the fact that it implies a certain minimal amount of differentiability of the invariant manifold. This is the content of the following result.

THEOREM 6. *Let $V \subset M$ be an r-normally hyperbolic invariant manifold for a C^k diffeomorphism φ, $k \geq 1$. If V is compact and C^1 and if $k \geq r$ then V is C^r, in the sense that it is locally the graph of a C^r function whose $[r]^{\text{th}}$ derivative is $(r - [r])$-Hölder.*

Here $[r]$ denotes the integer part of r. We observe that although in [HPS,1977] this theorem is stated only for r an integer (Theorem 4.1 on page 39), the general case follows from Remark 2 (on page 38) dealing with Hölder sections.

In this theorem we require V to be compact so it cannot be immediately applied to the locally invariant submanifolds that the previous theorem dealt with. Still, one can apply the same techniques in this situation to get the following result. Let $\varphi: M \to M$, $p \in M$ and $T_pM = E_1 \oplus E_2$ be as above. Take $r \geq 1$ such that for some $C > 0$ and some $\sigma > 1$ we have $\|(d\varphi)_p^n v_2\| / \|(d\varphi)_p^n v_1\|^r \geq C\sigma^n$ for all $n \geq 0$ and all $v_1 \in E_1$, $v_2 \in E_2$ with $\|v_1\| = 1 = \|v_2\|$. Suppose that $r \leq k$ and let V be a locally invariant submanifold with $T_pV = E_1$; then V is r-normally hyperbolic at the fixed point p. If $d\varphi_p \mid E_1$ has only contracting eigenvalues or if $d\varphi_p \mid E_1$ has non-contracting eigenvalues but we take $\varepsilon > 0$ small enough in the construction of V (recall the outline of this construction in Theorem 5), then V is C^r.

We use this to justify our claims about the differentiability of the invariant submanifolds introduced above. It is not difficult to check that V is r-normally hyperbolic at p if and only if

$$\frac{|\lambda_2|}{|\lambda_1|^r} > 1 \tag{1}$$

whenever λ_1, λ_2 are eigenvalues of $(d\varphi)_p$ with $|\lambda_1| < a < |\lambda_2|$. Let us consider first the centre-stable manifold. Clearly, in this case we may take r

arbitrarily large satisfying (1). Hence, if φ is C^k, k finite, then $W^{cs}(p)$ may be chosen C^k, as we claimed. We point out that this is not true for $k = \infty$: in general C^∞ diffeomorphisms do not have C^∞ centre-stable manifolds [S,1979], although they do have them of class C^k for any $k \geq 1$. This is related with the fact that $W^{cs}(p)$ is in general not unique. Analogous comments hold for the centre-unstable manifold and the centre manifold. Consider now the strong stable manifold. Again, we may take r as in (1) with arbitrarily large value, so $W^{ss}(p)$ is C^k if φ is C^k. But now, due in part to the uniqueness of $W^{ss}(p)$, this is still true for $k = \infty$. The same holds for the strong unstable manifold, the stable manifold and the unstable manifold.

We now want to use the theory of normal hyperbolicity to study the stable and unstable foliations, defined in a neighbourhood of a basic set of a surface diffeomorphism $\varphi \colon M \to M$. Here, we are mostly concerned with the differentiability of these foliations. We first treat in some detail the fixed point case and then show how the conclusions can be extended to general basic sets.

For diffeomorphisms $\varphi \colon M \to M$ as above, we construct a lift

$$
\begin{array}{ccc}
L(M) & \xrightarrow{\; D\varphi \;} & L(M) \\
\downarrow & & \downarrow \\
M & \xrightarrow{\; \varphi \;} & M
\end{array}
$$

where $L(M) = \{(x, L) \mid x \in M$ and L is a one-dimensional linear subspace of $T_x M\}$ and $D\varphi$ is the diffeomorphism induced on $L(M)$ by the derivative of φ. Clearly, if φ is C^k then $D\varphi$ is C^{k-1}. Let $\mathcal{F}$ be a foliation of an open subset U of M by differentiable curves. Then $\tilde{U} = \{(x, L) \mid x \in U$ and L is the tangent space of the $\mathcal{F}$-leaf through x$\}$ is a surface in $L(M)$ and $\mathcal{F}$ is a φ- invariant foliation if and only if $\tilde{U}$ is a $D\varphi$-invariant manifold. If $\tilde{U}$ is C^r then $\mathcal{F}$ is C^r (i.e. there are C^r local coordinates which trivialize $\mathcal{F}$) and we even have in this case that *the tangent spaces* $T_x\mathcal{F}(x)$ *to the leaves of* $\mathcal{F}$ *depend* C^r *on* x.

Let now φ be a C^2 diffeomorphism of $M = \mathbb{R}^2$ and $0 \in \mathbb{R}^2$ be a fixed point of φ, with $(d\varphi)_0 = \begin{pmatrix} \sigma & 0 \\ 0 & \lambda \end{pmatrix}$, $0 < |\sigma| < 1 < |\lambda|$. If E^u denotes the unstable eigenspace of $(d\varphi_0)$ then $O^u = (0, E^u) \in L(\mathbb{R}^2)$ is a fixed point of $D\varphi$. The three eigenvalues of $d(D\varphi)_{O^u}$ are $\sigma\lambda^{-1} < \sigma < 1 < \lambda$. Hence, there is a two-dimensional locally invariant C^1 manifold $\tilde{U}$ containing O^u, such that $T_{O^u}\tilde{U}$ is the eigenspace of $d(D\varphi)_{O^u}$ corresponding to the eigenvalues σ and λ. Note that $\tilde{U}$ projects diffeomorphically onto a neighbourhood U of $0 \in \mathbb{R}^2$. Then, there is a (unique) foliation $\mathcal{F}^u$ such that $T_x\mathcal{F}^u(x) = L$ for all $(x, L) \in \tilde{U}$. Clearly, $\mathcal{F}^u$ is a φ-invariant C^1 foliation and $\mathcal{F}^u(0) \subset W^u(0)$

(because $\mathcal{F}^u(0)$ is an invariant submanifold tangent to E^u at the origin), so $\mathcal{F}^u$ is an unstable foliation on a neighbourhood of $0 \in \mathbb{R}^2$. Moreover, one easily checks that $\tilde{U}$ is r-normally hyperbolic at O^u for all $r < 1 - \log \lambda / \log \sigma$. Hence, if φ is C^3 and $0 < \varepsilon < -\log \lambda / \log \sigma$, $\varepsilon \le 1$ then $\tilde{U}$ is $C^{1+\varepsilon}$, and so $\mathcal{F}^u$ is a $C^{1+\varepsilon}$ foliation. In the same way we construct a C^1 stable foliation $\mathcal{F}^s$ near zero and we show that $\mathcal{F}^s$ can be taken of class $C^{1+\varepsilon}$ if φ is C^3.

Before proceeding to the case of a general basic set, let us derive a corollary of the discussion above, which was much used in the text.

THEOREM 7 [**H**,1964]. *Let φ be a C^2 diffeomorphism of a surface M and $p \in M$ be a saddle point for φ. Then φ admits C^1-linearizing coordinates near p. If φ is C^3 such coordinates can even be taken to be $C^{1+\varepsilon}$ for some $\varepsilon > 0$.*

We detail the proof of the first part of the theorem; the C^3 case follows from the same argument. First we construct C^1-linearizing coordinates for $\varphi \mid W^s(p)$. Let $\alpha \colon \mathbb{R} \to W^s(p)$ be any C^2 parametrization of $W^s(p)$ with $\alpha(0) = p$, and define $f \colon \mathbb{R} \to \mathbb{R}$ by

$$f = \alpha^{-1} \circ \varphi \mid W^s(p) \circ \alpha.$$

Then $0 \in \mathbb{R}$ is an attracting fixed point for f with $f'(0) = \sigma$, the contracting eigenvalue of $(d\varphi)_p$. It is not difficult to check that $h(x) = \lim_{n \to \infty} \frac{f^n(x)}{\sigma^n}$ is a well defined C^1 function in a neighbourhood of 0 ([**S**,1957]). Clearly, we have $h(f(x)) = \sigma \cdot h(x)$ for all x near 0. It follows that $\xi = h \circ \alpha^{-1}(x)$ is a C^1-linearizing change of coordinates for $\varphi \mid W^s(p)$ near p, as we wished. Replacing φ by φ^{-1}, we get in the same way a C^1 coordinate, say η, defined on $W^u(p)$ (near p), relative to which $\varphi \mid W^u(p)(\eta) = \lambda \eta$, λ being the expanding eigenvalue of $(d\varphi)_p$. We now define coordinates (ξ, η) in a neighbourhood of p as follows. Let $\mathcal{F}^s$, $\mathcal{F}^u$ be the stable and unstable foliations constructed above. For $\xi \in W^s(p)$ and $\eta \in W^u(p)$ we take (ξ, η) to be the intersection point of $\mathcal{F}^u(\xi)$ and $\mathcal{F}^s(\eta)$. Since $\mathcal{F}^s$ and $\mathcal{F}^u$ are C^1 and transversal to each other this really defines C^1 coordinates near p. Moreover, due to the invariance of the foliations we have

$$\varphi(\xi, \eta) = (\varphi \mid W^s(p)(\xi),\ \varphi \mid W^u(p)(\eta)) = (\sigma \xi, \lambda \eta),$$

so the proof is complete.

Now we indicate how to obtain stable and unstable foliations in a neighbourhood of a basic set Λ of a two-dimensional diffeomorphism φ. One way that this can be done is by the limiting process described in Chapter 2 for the case of a horseshoe. A different but related proof involving the λ-lemma, which we outlined above for the case of a hyperbolic fixed point, can also be used to construct these foliations for any basic set of a surface diffeomorphism. Using this approach, we sketch the construction of the unstable foliation $\mathcal{F}^u$ and refer the reader to [**M**,1973a] for more details; of course, the stable foliation is obtained in a similar way. After this is done, we discuss the differentiability of the foliations when the diffeomorphism is of class C^2 or C^3. We observe that if the diffeomorphism is only C^1, as considered in [**M**,1973a], then the foliations are in general not differentiable.

To begin with define $\mathcal{F}^u$ in a fundamental neighbourhood of $W^s(\Lambda)$ that is in an open neighbourhood N of $W_\varepsilon^s(\Lambda) - \varphi(W_\varepsilon^s(\Lambda))$, small $\varepsilon > 0$, such that $N \cap \Lambda = \emptyset$. That such a neighbourhood exists results from the fact that Λ is isolated [**HPPS**,1970] (see also [**N**,1980], [**S**,1978]). The definition of $\mathcal{F}^u$ on N must be done in such a way that $\varphi(\mathcal{F}^u(x)) \supset \mathcal{F}^u(\varphi(x))$ for all $x \in N \cap \varphi^{-1}(N)$. Then we use the positive iterates of φ to extend $\mathcal{F}^u$ and finally we include the (local) unstable manifolds of the points in Λ as leaves of $\mathcal{F}^u$. One can show directly that since $\mathcal{F}^u$ has codimension 1 and we assume φ to be C^2, this procedure yields a C^1 foliation $\mathcal{F}^u$, which is the required unstable foliation. A main ingredient here is normal hyperbolicity of such foliation which we now further explore.

Let $\tilde{U}$ be the locally invariant submanifold of $L(M)$, defined by $\tilde{U} = \{(x, T_x\mathcal{F}^u(x)) \mid x \in U\}$, U being some neighbourhood of Λ where $\mathcal{F}^u$ is defined. We prove that $\tilde{U}$ is $(1 + \varepsilon)$-normally hyperbolic for some $\varepsilon > 0$, at least if U is small enough. We assume that on U there is also a stable foliation $\mathcal{F}^s$ (obtained by the dual construction). In order to describe the derivatives of φ and $D\varphi$ more explicitly, we take nonzero C^1 vector fields X^u and X^s, tangent to leaves of $\mathcal{F}^u$ respectively $\mathcal{F}^s$ and define the functions $\lambda^u, \lambda^s \colon U \to \mathbb{R}$ by

$$d\varphi(X^u(x)) = \lambda^u(x) \cdot X^u(\varphi(x)),$$
$$d\varphi(X^s(x)) = \lambda^s(x) \cdot X^s(\varphi(x)).$$

By a proper choice of X^u and X^s, we obtain that $|\lambda^u| > 1$ and $|\lambda^s| < 1$ on all of Λ and even on U if U is a sufficiently small neighbourhood of Λ. Next, we define bases in the tangent spaces of $L(M)$ at the points of $\tilde{U}$. We denote the point $(x, T_x\mathcal{F}^u(x)) \in \tilde{U}$ by x^u. The vectors $\tilde{X}^u(x^u)$, $\tilde{X}^s(x^u)$ in $T_{x^u}(L(M))$ are defined to be the tangent vectors of $\tilde{U}$ which project on $X^u(x)$ and $X^s(x)$. As a third basis vector in $T_{x^u}(L(M))$, we take the tangent vector of the curve

$$t \mapsto (x, L_t) \in L(M),$$

with $L_t \subset T_x(M)$ being the line spanned by $X^u(x) + t \cdot X^s(x)$. This basis vector is denoted by $Z(x^u)$. Then

$$
\begin{aligned}
d(D\varphi)(\tilde{X}^u(x^u)) &= \lambda^u(x) \cdot \tilde{X}^u(\varphi(x)^u), \\
d(D\varphi)(\tilde{X}^s(x^u)) &= \lambda^s(x) \cdot \tilde{X}^s(\varphi(x)^u), \\
d(D\varphi)(Z(x^u)) &= \lambda^s(x) \cdot (\lambda^u(x))^{-1} \cdot Z(\varphi(x)^u).
\end{aligned}
$$

So $\tilde{U}$ is normally contracting. There is an $\varepsilon > 0$ such that for each $x \in U$,

$$
|\lambda^s(x)|^{1+\varepsilon} \geq \theta \cdot |\lambda^s(x) \cdot (\lambda^u(x))^{-1}|
$$

for some fixed $\theta > 1$. Hence, $\tilde{U}$ is $(1 + \varepsilon)$-normally hyperbolic. Then, although $\tilde{U}$ is not compact and only locally invariant, still the ideas in the proof of the last theorem above can be applied to conclude that $\tilde{U}$ is $C^{1+\varepsilon}$, and so $\mathcal{F}^u$ is $C^{1+\varepsilon}$, if φ is C^3. As observed before, it even follows that the tangent spaces $T_x\mathcal{F}^u(x)$ vary $C^{1+\varepsilon}$ with x in U.

Differentiable dependence of invariant manifolds and foliations on diffeomorphisms. As we observed at various places, the invariant manifolds and foliations which we constructed in this appendix depend continuously on the diffeomorphism (as far as they are unique or else as far as they are obtained by a prescribed construction making them unique). Sometimes one needs more, and we describe here for diffeomorphisms φ_μ how these invariant manifolds and foliations depend differentiably on a parameter μ. We assume, here, that $\varphi_\mu(x)$ is a C^k function of (μ, x).

First, we discuss the case of the stable manifold of a hyperbolic fixed point. Let p_μ be a hyperbolic fixed point of φ_μ as above, depending continuously (and hence C^k, by the implicit function theorem) on μ. We let Σ be the parameter space and M the ambient manifold and consider the diffeomorphism $\Phi: \Sigma \times M \to \Sigma \times M$ defined by $\Phi(\mu, x) = (\mu, \varphi_\mu(x))$. Then $W = \bigcup_\mu (\{\mu\} \times W^s(p_\mu))$ is a centre-stable manifold for each of the fixed points (μ, p_μ) of Φ and

$$
W = \bigcup_{n \geq 0} \Phi^{-n}(V) \tag{2}
$$

for any neighbourhood V of $\{(\mu, p_\mu): \mu \in \Sigma\}$ in W. Actually, it is not difficult to check that in this context the centre-stable manifold is unique and may be obtained by a local construction as in the stable manifold theorem (there is no need for the globalizing procedure involving a bump-function). Therefore, W is a C^k injectively immersed submanifold of $\Sigma \times M$ and this is the sense in which we say that $W^s(p_\mu)$ depends differentiably on μ. Here Σ and M may be arbitrary Banach manifolds; an interesting case is $\Sigma =$ open subset of $\mathrm{Diff}^k(M)$. Observe also that the argument remains valid for $k = \infty$: if

$(\mu, x) \mapsto \varphi_\mu(x)$ is C^∞ then W is C^∞, as a consequence of uniqueness and property (2).

A similar conclusion may be obtained for stable manifolds of hyperbolic sets, by applying the construction above to the operator $F_\varphi(f) = \varphi f \varphi^{-1}$ introduced before. If Λ_μ is a hyperbolic set for φ_μ, depending continuously on μ and such that it is the maximal invariant set in some neighbourhood, then, as we have seen, there are conjugacies $h_{\mu,\nu}: \Lambda_\mu \to \Lambda_\nu$ between $\varphi_\mu \mid \Lambda_\mu$ and $\varphi_\nu \mid \Lambda_\nu$, depending continuously on (μ, ν). This $h_{\mu,\nu}$ is a hyperbolic fixed point of $G_{\mu,\nu}: C(\Lambda_\mu, M) \hookleftarrow$, $G_{\mu,\nu}(f) = \varphi_\mu f \varphi_\nu$ and $h_{\mu,\mu} = i$ is just the inclusion of Λ_μ in M. Then, $W = \bigcup_\nu W^s(h_{\mu,\nu})$ is a C^k centre-stable manifold for the fixed point (μ, i) of $(\nu, f) \mapsto (\nu, G_{\mu,\nu}(f))$ and, so, for each $x \in \Lambda_\mu$, $\bigcup_\nu(\{\nu\} \times W^s(h_{\mu,\nu}(x))) = ev_x(W)$ is a C^k submanifold of $\Sigma \times M$.

We now state a result concerning differentiability on parameters of the foliations themselves, *specially for stable and unstable foliations defined in a neighbourhood of basic sets in two dimensions*. Let U be a small neigbourhood of a basic set $\Lambda_\mu \subset M^2$. With the same notations as above we have the following result.

THEOREM 8. *If $\Phi: \Sigma \times M \to \Sigma \times M$ is C^2 then there are transverse invariant foliations $\mathcal{F}_\mu^s(x), \mathcal{F}_\mu^u(x)$ defined on U, such that the maps $(\mu, x) \mapsto T_x \mathcal{F}_\mu^s(x)$ and $(\mu, x) \mapsto T_x \mathcal{F}_\mu^u(x)$ are C^1. Moreover they are $C^{1+\varepsilon}$ if Φ is C^3, for some $\varepsilon > 0$.*

The construction of $\mathcal{F}_\mu^s, \mathcal{F}_\mu^u$ is a parametrized version of the procedure described above for neighbourhoods of basic sets in two dimensions.

The conclusion that $\mathcal{F}_\mu^s(x)$ and $\mathcal{F}_\mu^u(x)$ are differentiable on (μ, x) is again related to the fact that the invariant submanifolds

$$\bigcup_\mu(\{\mu\} \times \{(x, T_x\mathcal{F}_\mu^s(x)) \mid x \in U\})$$

and

$$\bigcup_\mu(\{\mu\} \times \{(x, T_x\mathcal{F}_\mu^u(x)) \mid x \in U\})$$

are $(1 + \varepsilon)$-normally hyperbolic.

Endomorphisms. So far, φ was always supposed to be a diffeomorphism of M. For some applications (e.g. Appendix 4), one needs versions of the results and constructions for general (possibly noninvertible) smooth maps. Here, we let φ be such a map, of class C^k, and explain how a good deal of what has been said applies to hyperbolic periodic points or invariant sets of φ.

Let first p be a hyperbolic fixed (or periodic) point: all eigenvalues of $d\varphi_p$ have norm different from 1 (zero is allowed). The stable manifold of p is defined as for diffeomorphisms and the unstable manifold is the set of points x_0 such that there exists a sequence $(x_{-n})_{n\geq 0} \to p$ with $\varphi(x_{-n}) = x_{-n+1}$ for all $n \geq 1$. Local unstable manifolds (meaning small compact neighbourhoods of p in $W^u(p)$) may be constructed by a graph transform procedure, in essentially the same way as in the invertible case. This is also possible for local stable manifolds, although it requires more care: one bypasses the lack of the inverse map by defining the graph transform through an implicit function argument; see Ex. III.2 in [**S**,1978]. Alternatively, one may construct these local stable manifolds by Perron's method, which does not involve φ^{-1}; see [**PM**,1982], [**S**,1978]. Both local stable and local unstable manifolds are C^k (C^∞ if φ is C^∞) embedded disks tangent to the stable and unstable eigenspaces of $d\varphi_p$. They depend continuously (and even differentiably, see above) on the map in the C^k topology. Clearly, this means that compact parts of $W^u(p)$ (i.e. the positive iterates of the local unstable manifold) vary continuously with φ. For the stable manifold the situation is more subtle: $W^s(p)$ is the union of the pre-images of any local stable manifold but this may not be a submanifold (not even connected). Continuity of the stable manifold in a global way is discussed in Appendix 4.

We now deal with the more general case of a hyperbolic set of an endomorphism. A compact set $\Lambda \subset M$ with $\varphi(\Lambda) = \Lambda$ is a hyperbolic set for φ if there are $C > 0$, $0 < \lambda < 1$ and a continuous splitting $T_\Lambda M = E^s \oplus E^u$ of its tangent bundle, with

$$d\varphi(E^s) \subset E^s \text{ and } \|d\varphi^n \,|\, E^s\| \leq C\lambda^n \quad \forall\, n \geq 1,$$
$$d\varphi(E^u) = E^u, \ d\varphi \,|\, E^u \text{ invertible and } \|d\varphi^{-n} \,|\, E^u\| \leq C\lambda^n, \quad \forall\, n \geq 1.$$

It is convenient at this point to introduce the projective limit of φ. This is the invertible map $\hat{\varphi}: \hat{M} \to \hat{M}$ defined by $\hat{\varphi}(\{x_k\}) = \{\varphi(x_k)\}$ with

$$\hat{M} = \hat{M}(\varphi) = \{\{x_k\} \in M^{\mathbf{Z}} \,|\, x_{k+1} = \varphi(x_k), \ \forall\, k \in \mathbf{Z}\}.$$

Note that $\pi \circ \hat{\varphi} = \varphi \circ \pi$ where $\pi: \hat{M} \to M$ is given by $\pi(\{x_k\}) = x_0$. Local stable and unstable manifolds of points $\hat{x} = \{x_k\} \in \hat{\Lambda} = \hat{M} \cap \Lambda^{\mathbf{Z}}$ are defined by

$$W^s_\varepsilon(\hat{x}) = \{\{y_k\} \in \hat{M} \,|\, \text{dist}\,(y_k, x_k) \leq \varepsilon, \ \forall\, k \geq 0\},$$

and dually for $W^u_\varepsilon(\hat{x})$. As before, we actually have that dist $(y_k, x_k) \to 0$ as $k \to +\infty$, for $\{y_k\}$ in $W^s_\varepsilon(\hat{x})$. They project under π onto embedded C^k disks in M tangent to $E^s(x_0)$ and $E^u(x_0)$, respectively. *We point out that $\pi(W^s_\varepsilon(\hat{x}))$ is completely determined by $x_0 = \pi(\hat{x})$, whereas $\pi(W^u_\varepsilon(\hat{x}))$ in general depends on the whole negative orbit $(x_k)_{k\leq 0}$.* Hyperbolicity of

Λ translates into hyperbolicity of $\pi \,|\, \hat{\Lambda}$ as a fixed point of the operator $F_\varphi \colon C(\hat{\Lambda}, M) \hookleftarrow$, $F_\varphi(g) = \varphi \circ g \circ (\hat{\varphi})^{-1}$. Then, for ψ close to φ, there is $h_\psi \colon \hat{\Lambda} \to M$ close to π such that $\psi \circ h_\psi = h_\psi \circ \hat{\varphi}$. The set $\Lambda_\psi = h_\psi(\hat{\Lambda})$ is hyperbolic for ψ and h_ψ lifts to a conjugacy $\hat{h}_\psi \colon \hat{\Lambda} \to \hat{\Lambda}_\psi$ between $\hat{\varphi} \,|\, \hat{\Lambda}$ and $\hat{\psi} \,|\, \hat{\Lambda}_\psi$. Moreover, $\pi(W^s_\varepsilon(h_\Psi(\hat{x})))$ is uniformly C^k-close to $\pi(W^s_\varepsilon(\hat{x}))$ for all $\hat{x} \in \hat{\Lambda}$ and an analogous statement holds for unstable manifolds. In general $\varphi \,|\, \Lambda$ and $\psi \,|\, \Lambda_\psi$ are not conjugate, but in the case we are most interest in, namely the quadratic endomorphisms of Chapter 6 (see also Appendix 4), we have at least a semi-conjugacy from $\psi \,|\, \Lambda_\psi$ to $\varphi \,|\, \Lambda$. More precisely, if Λ is a hyperbolic set of an endomorphism $\varphi(x, y) = (y, y^2 + \mu)$ of $\mathbb{R}^2$ and ψ is close to φ, then $H_\psi \circ h_\psi = \pi$ defines a continuous surjection $H_\psi \colon \Lambda_\psi \to \Lambda$ which satisfies $\varphi \circ H_\psi = H_\psi \circ \psi$.

This can be seen by noting that if $\{x_k\}$ and $\{y_k\}$ have the same image under h_ψ, then x_0 must belong to the local stable manifold of y_0 and, thus, $x_0 = y_0$. It is also important to observe that in this case, $\pi(W^u_\varepsilon(\hat{x}))$ depends only on x_0: for any $\hat{y} = \{y_k\}$ with $y_0 = x_0$ we have that $\pi(W^u_\varepsilon(\hat{y}))$ and $\pi(W^u_\varepsilon(\hat{x}))$ represent the same segment in the parabola (actually the end points may vary slightly, but this is irrelevant for our purposes).

APPENDIX 2

MARKOV PARTITIONS

In Chapter 4 we mentioned the existence of Markov partitions for dynamically defined Cantor sets and their relation with Markov partitions for hyperbolic diffeomorphisms in two dimensions. Here, we shall discuss these points further: we construct partitions for basic sets of surface diffeomorphisms and apply them to get the corresponding structure for Cantor sets. We observe that Markov partitions also exist in higher dimensions. However, as we explain below, in two dimensions their construction is simpler (and they have nicer properties).

Let $\varphi \colon M \to M$ be a diffeomorphism on a compact 2-manifold and let Λ be a basic set for φ. That is, Λ is a compact, invariant, hyperbolic set which is transitive and contains a dense subset of periodic orbits; moreover, Λ has local product structure or, equivalently, Λ is the maximal invariant set in some neighbourhood of it (see Chapter 0). We suppose that Λ is not just a periodic orbit, in which case we say that Λ is *trivial*. There are three possibilities for Λ: either Λ is all of M and then M is a torus and φ is an Anosov diffeomorphism, or Λ is an attractor (repeller) or Λ is of saddle-type. The second possibility, Λ a hyperbolic attractor, which is locally the product of an interval and a Cantor set, includes Plykin's attractors on S^2 (see [**GH**,1983]) and the 1-dimensional attractors in the DA (derived from Anosov) maps defined by Smale (see [**S**,1967],[**W**,1970]). Finally, the third possibility includes the horseshoes constructed in Chapter 2; in our context of homoclinic bifurcations, this is the most relevant case. Also, in what follows, we are going to assume that Λ is *topologically mixing*; this means that for any two open sets U, V, in Λ, we have $\varphi^n(U) \cap V \neq \phi$ for all sufficiently big n; this implies that all nonzero powers of φ are topologically transitive. Our assumption is justified by the fact that we can always decompose Λ into a finite union of topologically mixing components for some power φ^m of φ, as can easily be shown.

For $x \in \Lambda$ we define local stable and unstable manifolds as in Appendix 1:

$$W_\varepsilon^u(x) = \{y \in M | \lim_{n \to -\infty} \rho(\varphi^n(x), \varphi^n(y)) = 0 \quad \text{and}$$
$$\text{for all } n \leq 0, \rho(\varphi^n(x), \varphi^n(y)) \leq \varepsilon\},$$
$$W_\varepsilon^s(x) = \{y \in M | \lim_{n \to +\infty} \rho(\varphi^n(x), \varphi^n(y)) = 0 \quad \text{and}$$
$$\text{for all } n \geq 0, \rho(\varphi^n(x), \varphi^n(y)) \leq \varepsilon\},$$

where ρ denotes the distance with respect to some fixed Riemannian metric.

From the local product structure, we know that for $x, x' \in \Lambda$ sufficiently close, $W_\varepsilon^u(x)$ and $W_\varepsilon^s(x')$ have a unique point of intersection and that this point also belongs to Λ.

We say that x is a *boundary point* of Λ in the *unstable direction* if x is a boundary point of $W_\varepsilon^u(x) \cap \Lambda$, i.e. if x is an accumulation point only from one side by points in $W_\varepsilon^u(x) \cap \Lambda$. If x is a boundary point of Λ in the unstable direction, then, due to the local product structure, the same holds for all points in $W_\varepsilon^s(x) \cap \Lambda$. So the boundary points in the *unstable* direction are locally intersections of local *stable* manifolds with Λ. For this reason we denote the set of boundary points in the unstable direction by $\partial_s \Lambda$. The boundary points in the stable direction are defined similarly; the set of these boundary points is denoted by $\partial_u \Lambda$. Notice that if $\Lambda = M^2$ then $\partial_s \Lambda = \partial_u \Lambda = \phi$ and if Λ is a one-dimensional attractor then $\partial_s \Lambda = \phi \neq \partial_u \Lambda$.

Our construction of Markov partitions for Λ is based on the following simple theorem from [**NP**,1973].

THEOREM 1. *For a basic set Λ as above there is a finite number of (periodic) saddle points $p_1^s, \cdots, p_{n_s}^s$ such that*

$$\Lambda \cap \left(\bigcup_i W^s(p_i^s) \right) = \partial_s \Lambda.$$

Similarly, there is a finite number of (periodic) saddle points $p_1^u, \cdots, p_{n_u}^u$ such that

$$\Lambda \cap \left(\bigcup_i W^u(p_i^u) \right) = \partial_u \Lambda.$$

PROOF: Let $x \in \partial_s \Lambda$. We first claim that $W^s(x)$ is periodic, i.e. $x \in W^s(p)$ for some periodic point $p \in \Lambda$. If we suppose that this is not the case we reach a contradiction. In fact, let $\varphi^{n_i}(x)$ be a converging sequence of iterates of x, $n_i \to \infty$. Since $W^s(x)$ is not periodic, $\varphi^{n_i}(x)$ and $\varphi^{n_j}(x)$ are in different stable manifolds if $n_i \neq n_j$. But, by local product structure (see Chapter 0), there is $N > 0$ such that for $n_i, n_j > N$ and $n_i \neq n_j$, $W_\varepsilon^u(\varphi^{n_i}(x)) \cap W_\varepsilon^s(\varphi^{n_j}(x))$ consists of exactly one point and this point belongs to Λ. It then follows that there are arbitrarily large $n_i, n_j, n_\ell > N$ such that the points $W_\varepsilon^u(\varphi^{n_i}(x)) \cap W_\varepsilon^s(\varphi^{n_j}(x))$ and $W_\varepsilon^u(\varphi^{n_i}(x)) \cap W_\varepsilon^s(\varphi^{n_\ell}(x))$ are on different sides of $\varphi^{n_i}(x)$ in $W_\varepsilon^u(\varphi^{n_i}(x))$. Since φ^{-n} decreases distances along unstable manifolds, we see that x is accumulated from both sides in $W^u(x)$ by points of $W^u(x) \cap \Lambda$, which is a contradiction. (Observe that the same reasoning holds in higher dimensions if the unstable bundle of Λ has codimension 1). The argument for points in $\partial_u \Lambda$ is similar. We now claim that if p is periodic and $W^s(p) \cap \partial_s \Lambda \neq \phi$, then $W^s(p) \cap \Lambda \subset \partial_s \Lambda$. Indeed, since all sets involved in the claim are invariant under φ, we can localize the question near p. But we already observed that if $x \in \partial_s \Lambda$ then the same is true for all points in its local stable manifold. This shows that $\partial_s \Lambda = \Lambda \cap \left(\bigcup_i W^s(p_i^s) \right)$, where each p_i is a periodic point. The proof that there are only finitely many such periodic points p_i^s is now clear from the arguments above. In fact, if this were not the case, one would just take an accumulation point say y of these periodic points. Now, by local product structure, the local unstable manifold of y intersects many $W^s(p_i^s)$ in points of Λ and this shows that not all $W^s(p_i^s)$ bound Λ. Thus, we reach a contradiction and the proof of the theorem is complete. □

We note that this theorem, which is typically two-dimensional, makes the construction of Markov partitions in two dimensions simpler than in higher dimensions. Another reason why the two-dimensional case is easier is that in this dimension one can make boxes (see below), whose boundaries consist of pieces of stable and unstable separatrices. For the general construction see Bowen [**B**,1975a].

Now we come to the *definition* of a *Markov partition* for a *basic set* Λ as introduced above. Such a Markov partition consists of a finite set of *boxes*, i.e. diffeomorphic images of the square $Q = [-1, +1]^2$, say $B_1 = \Psi_1(Q), \cdots, B_\ell = \Psi_\ell(Q)$ such that

(i) $\Lambda \subset \bigcup_i B_i$,

(ii) $\overset{\circ}{B}_i \cap \overset{\circ}{B}_j = \phi$, $i \neq j$, where $\overset{\circ}{B}$ denotes the interior of B,

(iii) $\varphi(\partial_s B_i) \subset \bigcup_j \partial_s B_j$ and

$\varphi^{-1}(\partial_u B_i) \subset \bigcup_j \partial_u B_j$, where

$\partial_s B_i = \Psi_i(\{(x,y) | -1 \leq x \leq 1, |y| = 1\})$ and
$\partial_u B_i = \Psi_i(\{(x,y) | |x| = 1, -1 \leq y \leq 1\})$,

(iv) there is a positive integer n such that $\varphi^n(B_i) \cap B_j \neq \phi$ for all $1 \leq i, j \leq \ell$
(this means that the Markov partition is mixing).

The geometric consequences of the third condition are indicated in Figure A2.1.

Usually one also requires that $\varphi(B_i) \cap B_j$ is either empty or connected. For our present considerations this is not important, but one can always satisfy this last condition by taking the boxes of the Markov partition sufficiently small.

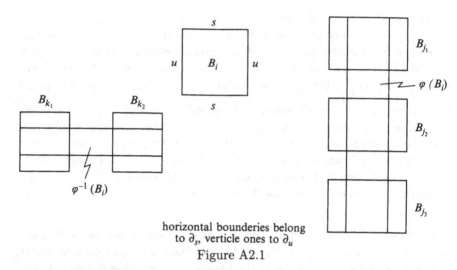

horizontal bounderies belong
to ∂_s, verticle ones to ∂_u

Figure A2.1

THEOREM 2. *There is a Markov partition for Λ with arbitrarily small diameter.*

PROOF: Let us begin the construction of a Markov partition in case $\partial_s \Lambda \neq \phi \neq \partial_u \Lambda$. For this we take closed arcs

$$I_1^s, \cdots, I_{n_s}^s \text{ in } W^s(p_1^s), \cdots, W^s(p_{n_s}^s) \text{ and}$$
$$I_1^u, \cdots, I_{n_u}^u \text{ in } W^u(p_1^u), \cdots, W^u(p_{n_u}^u),$$

where $p_1^s, \cdots, p_{n_s}^s$ and $p_1^u, \cdots, p_{n_u}^u$ are (periodic) saddle points as in the above theorem. These intervals are chosen so that for each i, $\partial I_i^s \subset \bigcup_j I_j^u$ and

$\partial I_i^u \subset \bigcup_j I_j^s$. Moreover, if $p_i^s \notin \partial_u \Lambda$ then p_i^s is contained in the interior

of I_i^s; if $p_i^s = p_j^u \in \partial_s \Lambda \cap \partial_u \Lambda$ then this is a boundary point of both I_i^s and I_j^u. We assume, except for this last case, that the intervals I_i^s and I_j^u have disjoint boundaries. It is possible to satisfy these conditions since

both $\left(\bigcup_j W^u(p_j^u)\right) \cap \Lambda$ and $\left(\bigcup_j W^s(p_j^s)\right) \cap \Lambda$ are dense in Λ. (For each

periodic point $p \in \Lambda$, the components of $W^s(p) - \{p\}$ and $W^u(p) - \{p\}$ are dense in Λ, or disjoint from Λ, since we assume φ to be mixing on Λ.)

We shall prove that, if the arcs $I_1^s, \cdots, I_{n_s}^s$ and $I_1^u, \cdots, I_{n_u}^u$ are sufficiently long, then they "divide" Λ according to a Markov partition with an arbitrarily small diameter. To be more precise, we fix $\varepsilon > 0$ sufficiently small and say that $x \in \Lambda$ is *ε-enclosed* by the above arcs if

$$W_\varepsilon^s(x) \cap \left(\bigcup_j I_j^u\right) \text{ contains } x \text{ or contains points on both sides of } x,$$

and if

$$W_\varepsilon^u(x) \cap \left(\bigcup_j I_j^s \right)$$ contains x or contains points on both sides of x.

We claim that if all I_j^u and I_j^s are extended at least to length ℓ from the saddle points p_j^u, p_j^s along $W^u(p_j^u), W^s(p_j^s)$ in both directions (as long as the corresponding branch of the separatrix contains points of Λ), then, for ℓ sufficiently big, they enclose all points of Λ.

We prove this claim by contradiction. Let x_i be a point of Λ which is not yet enclosed when $\ell = i$ for $i \in N$. By compactness, $\{x_i\}$ has an accumulation point, say $\bar{x}$. We show that for some finite ℓ_0 there is a neighbourhood of $\bar{x}$ such that all of its points are enclosed whenever $\ell \geq \ell_0$. We have to distinguish between $\bar{x} \in \partial_u\Lambda$ and $\bar{x} \notin \partial_u\Lambda$ and also between $\bar{x} \in \partial_s\Lambda$ and $\bar{x} \notin \partial_x\Lambda$. We consider the case $\bar{x} \in \partial_u\Lambda$, $\bar{x} \notin \partial_s\Lambda$; the other cases can be treated similarly.

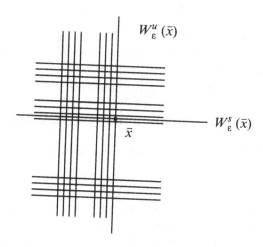

Figure A2.2

In Figure A2.2 we indicate local stable and unstable manifolds of points $x \in \Lambda$ near $\bar{x}$. Since $\bar{x} \in \partial_u\Lambda$, $\bar{x} \in \bigcup_j W^u(p_j^u)$ then, for some ℓ_1 and $\ell \geq \ell_1$, $\bar{x}$ belongs to $\bigcup_j I_j^u$. Since $\left(\bigcup_j W^u(p_j^u) \right) \cap \Lambda$ and $\left(\bigcup_j W^s(p_j^s) \right) \cap \Lambda$ are both dense in Λ, for some ℓ_2 and $\ell \geq \ell_2$, $\bigcup_j I_j^s$ and $\bigcup_j I_j^u$ contain segments I_j^s, $I_{j'}^s, I_{j''}^u$ as indicated in Figure A2.3.

For all $\ell \geq \ell_0 = \max(\ell_1, \ell_2)$, there is clearly a full neighbourhood of $\bar{x}$ in which all the points are enclosed. This gives the required contradiction and completes the proof of the claim.

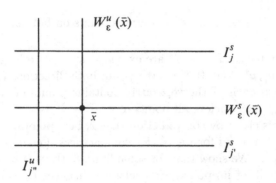

Figure A2.3

From now on we assume that the arcs I_j^s and I_j^u are so long that all points of Λ are ε-enclosed. For $\Lambda - \left(\bigcup_j \left(I_j^s \cup I_j^u \right) \right)$, we define an equivalence relation: $x \sim x'$ if we can join x by x' without crossing any of the arcs I_j^s or I_j^u. Now, it is easy to see that one can construct for each equivalence class a box, containing that equivalence class in its interior and whose boundary consists of segments of $\bigcup_j I_j^s$ and of $\bigcup_j I_j^u$ and such that the interior of the box contains no points of $\bigcup_j I_j^s$ or of $\bigcup_j I_j^u$. These boxes, one for each equivalence class, form a Markov partition. Notice that $\varphi(\partial_s B_i) \subset \bigcup_j \partial_s B_j$ because $\varphi\left(\bigcup_j I_j^s \right) \subset \bigcup_j I_j^s$, and $\varphi^{-1}(\partial_u B_i) \subset \bigcup_j \partial_u B_j$ because $\varphi^{-1}\left(\bigcup_j I_j^u \right) \subset \bigcup_j I_j^u$. Also, for φ small, the Markov partition has small diameter.

To conclude the proof of the theorem, we have now to consider the cases $\partial_u \Lambda = \partial_s \Lambda = \phi$ (Anosov case) and $\partial_u \Lambda \neq \phi = \partial_s(\Lambda)$ (one-dimensional attractors). In the first case, we take closed intervals I^s and I^u contained in $W^s(p)$ and $W^u(p)$, respectively, where p is a fixed (periodic) point. Again, if these intervals are long enough they ε-enclose all points in M and form a Markov partition with small diameter. In the remaining case we just consider the points $p_1^s, \cdots, p_{n_s}^s$ and closed intervals $I_1^s, \cdots, I_{n_s}^s$ together with a unique interval I^u in the unstable manifold of a periodic point and then repeat the construction above.   □

REMARK 1: In the case of a basic set of *saddle type*, we observe that although points of Λ may lie in the common boundary of two boxes, there is for each $x \in \Lambda$ a unique box B such that $x \in (\overset{o}{B} \cap \Lambda)$. In fact, in this case we can obtain a Markov partition with *all the boxes disjoint* by replacing each box B by the smallest box $\widetilde{B}$ such that

– $\widetilde{B} \cap \Lambda = C\ell \, (\overset{\circ}{\widetilde{B}} \cap \Lambda)$,

– $\partial \widetilde{B}$ consists of parts of local stable and local unstable manifolds of the periodic points $p_1^s, \cdots, p_{n_s}^s$ and $p_1^u, \cdots, p_{n_u}^u$.

REMARK 2: For a diffeomorphism C^r near φ, our construction yields nearby Markov partitions for the corresponding nearby basic set. This is due to the following facts. First of all, periodic points close to $p_1^s, \cdots, p_{n_s}^s$ and $p_1^u, \cdots, p_{n_u}^u$ have their stable and unstable manifolds bounding the nearby basic set. Secondly, on compact parts these manifolds are C^r-close to the original ones for φ.

Now we show how to construct Markov partitions for dynamically defined Cantor sets induced by basic sets as in Chapter 4. We choose a saddle point $p \in \Lambda$ and consider $W^s(p) \cap \Lambda$, or, more precisely for an arc I^s in $W^s(p)$ we consider $I^s \cap \Lambda$. We take I^s so long that it passes through all the boxes of the Markov partition. So, for each box B, $B \cap I^s$ consists of a number (at least one) of arcs in the s-direction passing from one component of $\partial_u B$ to the other component of $\partial_u B$. (See Figure A2.4.)

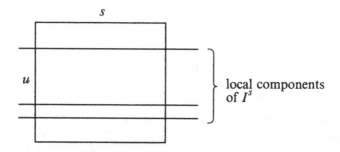

Figure A2.4

This fact, that one box B can be crossed several times by I^s, is incovenient because it makes the definition of a projection of Λ into $\Lambda \cap I^s$ ambiguous. For this reason we shall describe how to refine the Markov partition for a fixed I^s, so as to obtain a new Markov partition in which each box is crossed exactly once by I^s.

Let $I_1, I_2 \subset I^s$ be components of $B \cap I^s$. Let $\varphi(I_1)$ and $\varphi(I_2)$ be contained in the boxes B_1 and B_2; the component of $I_s \cap B_i$, containing $\varphi(I_i)$ is denoted by $\widetilde{I}_i$. If $B_1 = B_2$, we start again with $B_1 = B_2$ instead of B and $\widetilde{I}_i$ instead of I_i. We repeat this until we get $B_1 \neq B_2$: this must finally happen since under positive iterations of φ, I_1 and I_2 get more and more separated in the u-direction.

We have now the situation in Figure A2.5.

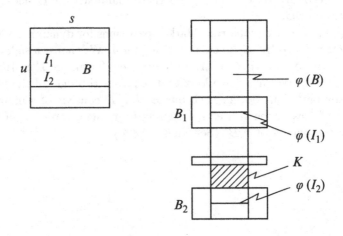

Figure A2.5

The region K of $\varphi(B)$, indicated in Figure A2.5 is just a connected component of $\varphi(B) - \left(\bigcup_j B_j \right)$, where $\bigcup_j B_j$ is the union of all the boxes of the Markov partition. Now we refine our Markov partition by removing from B the strip $\varphi^{-1}(K)$, thus splitting B in two smaller boxes. It is not hard to see that we obtain in this way a new partition in which I_1 and I_2 are not in the same box any more. So we diminished the set of pairs of components of $I^s \cap \left(\bigcup_j B_j \right)$ by one. Repeating this construction sufficiently often we obtain a Markov partition in which each box is crossed exactly once by I^s.

Next, we define a projection $\pi: \Lambda \to \Lambda \cap I^s$ by taking in each box the projection along local unstable manifolds into the intersection of I^s with that box. Then we define the expanding map $\Psi: I^s \cap \Lambda \to I^s \cap \Lambda$ as $\Psi = \pi \circ \varphi^{-N}$ (this is the same as in Chapter 4 except for the parametrization $\alpha: \mathbb{R} \to W^s(p)$). Extending the projection π from Λ to the union of the boxes of Λ, we get the map Ψ defined on a set of intervals $\{K_1, \cdots, K_k\}$ in I^s; each interval is mapped by Ψ diffeomorphically onto one of the intersections of I^s with a box of the Markov partition of Λ. The intervals $K_1, \cdots, K_k$ form the Markov partition of $I^s \cap \Lambda$ with expanding map Ψ.

From these constructions, it is clear that indeed the ideas of the example in Chapter 4, namely the construction of Markov partitions when the basic set is a horseshoe, carry over to the general situation of a two-dimensional basic set.

APPENDIX 3

ON THE SHAPE OF SOME
STRANGE ATTRACTORS

It is well known from numerical examples provided by Hénon [H,1976] that "strange attractors" in the plane may have the shape of (the closure of) the unstable separatrix of a saddle point with associated homoclinic intersections (i.e. the unstable separatrix crosses the local stable separatrix from one side to the other). Much attention has been given to the understanding of this phenomenon. Also, the "strange attractor" may disappear when the diffeomorphism is perturbed (without destroying the homoclinic intersections). We have already discussed in Chapter 7 the recent important results of Benedicks and Carleson and of Mora and Viana in this direction. Here, we present a previous partial explanation that has been provided independently by B. Szewc and F. Tangerman around 1981 but was never published. Their argument, in the simplest form, leads to the following proposition.

PROPOSITION 1. *Let $\varphi: \mathbb{R}^2 \to \mathbb{R}^2$ be a diffeomorphism with a hyperbolic fixed (periodic) saddle point p such that*

  – *$W^u(p)$ and $W^s(p)$ have a homoclinic intersection q, i.e. $p \neq q$,*

  – *the norm of $\det(d\varphi)$ is everywhere smaller than 1,*

  – *$W^u(p)$ remains in a bounded region of $\mathbb{R}^2$.*

Then there is a nonempty open set $U \subset \mathbb{R}^2$, such that for each $x \in U$, $\omega(x) \subset \overline{W^u(p)}$, i.e. the distance from $\varphi^n(x)$ to $W^u(p)$ goes to zero for n going to infinity.

PROOF: Let $U \subset \mathbb{R}^2$ be a bounded open subset whose boundary consists of segments of $W^u(p)$ and $W^s(p)$. Such U exist: due to the homoclinic intersection q one can form a closed curve consisting of segments of $W^u(p)$ and $W^s(p)$.

Consider $\varphi^n(U)$ for $n \geq 0$. Its boundary consists also of segments of $W^s(p)$ and $W^u(p)$; the segments of $W^s(p)$ become shorter and converge to p for increasing n. From this, and the boundedness of $W^u(p)$, it follows that the boundary of $\varphi^n(U)$, and hence $\varphi^n(U)$, stays in a bounded part of $\mathbb{R}^2$. This means that $|\det d\varphi| < 1 - \varepsilon$ on $\bigcup_{n \geq 0} \varphi^n(U)$, for some $\varepsilon > 0$, and, hence, that the area of $\varphi^n(U)$ goes to zero as n increases.

Consequently, for any $x \in U$ and $n > 0$, n big, $\varphi^n(x)$ is near the boundary of $\varphi^n(U)$ and, hence, near $W^u(p) \cup W^s(p)$. Since the part of the boundary formed by $W^s(p)$ has decreasing length, $\varphi^n(x)$ is close to $W^u(p)$. This means that $\omega(x) \subset \overline{W^u(p)}$ and so the proposition is proved. $\qquad \square$

Note that this proposition explains that certain attractors are *contained in* $\overline{W^u(p)}$, but not that they fill out all of it. In fact, there could be only one periodic attractor in $\overline{W^u(p)}$. This is the reason why one is interested in nonexistence results for periodic attractors proved by Benedicks and Carleson [**BC**,1991].

It is not hard to see that the above proposition applies to the Hénon map $H: \mathbb{R}^2 \to \mathbb{R}^2$ given by $H(x,y) = \left(1 - \frac{7}{5}x^2 + y, \frac{3}{10}x\right)$; the existence of a homoclinic intersection was proved in [**MS**,1980]. In fact both Szewc and Tangerman proved more, namely that one can, in the case of the Hénon map, choose U to be a neighbourhood of $\overline{W^u(p)}$.

Another interesting case can be obtained from the second example in Chapter 1, the pendulum, through small perturbations. With a first perturbation, we make the diffeomorphism, or even the differential equation attracting towards the $\{E = 1\}$ level without perturbing the dynamics inside this level. With a second perturbation, we make transverse homoclinic intersections in all the branches of the separatrices of the saddle point $(\pi, 0)$ (and hence destroy $\{E = 1\}$ as an invariant set). Although this example is defined not on the plane but on an annulus, the arguments of the proof of the proposition above still work.

APPENDIX 4

INFINITELY MANY SINKS IN
ONE-PARAMETER FAMILIES
OF DIFFEOMORPHISMS

In this appendix we shall prove a refinement of the main results of Chapter 6.

In that chapter we considered one diffeomorphism φ with a homoclinic tangency, and then proved that near this diffeomorphism there are many diffeomorphisms with infinitely many sinks. Here, we consider a one-parameter family of diffeomorphisms φ_μ with a homoclinic tangency (for $\mu = 0$), which satisfies certain generic properties, and then show that *within this one-parameter family* there are many diffeomorphisms φ_μ, μ near zero, with infinitely many sinks (or sources). Such a result, based on Newhouse's original work, was obtained in [**R**,1983].

We first point out that we can get this result by a rather formal argument in the following way. Let $\mathcal{D} = \text{Diff}^2(M)$ be the space of C^2 diffeomorphsms on a 2-manifold M. Let $\Sigma \subset \mathcal{D}$ be a codimension-1 submanifold consisting of diffeomorphisms with homoclinic tangency. We assume the usual generic properties (quadratic tangency and $|\det d\varphi|$ at the saddle point different from 1) to be satisfied for all diffeomorphisms $\varphi \in \Sigma$. We shall consider one-parameter families $\mu \mapsto \varphi_\mu$ which intersect Σ transversally for $\mu = 0$. First, however, we reformulate the main result of Chapter 6.

Let $\mathcal{U} \subset \mathcal{D}$ be the (open) set of those diffeomorphisms φ which have a neighbourhood V such that, for each integer m, there is an open and dense subset $V^{(m)} \subset V$ so that each $\varphi' \in V^{(m)}$ has at least m periodic sinks (or sources). From the results in Chapter 6, it follows that $\Sigma \subset \overline{\mathcal{U}}$. This means that, for any integer n, the following property is generic for one-parameter families φ_μ as above: there is a parameter value $\tilde{\mu}$, with $|\tilde{\mu}| < 1/n$, which is contained in some interval $I \ni \tilde{\mu}$, such that for generic $\mu' \in I$, $\varphi_{\mu'}$ has infinitely many sinks (or sources). Hence generic one-parameter families as above have this property for all integers n.

The result which we want to prove here differs from the above one in a sharp sense: we shall show that the same conclusion holds for an *open* and *dense* subset (in the C^∞ topology) of such one-parameter families. Also we shall explicitly describe this open and dense set.

We consider one-parameter families $\varphi_\mu : M \to M$, M a 2-manifold, which are C^∞ in the sense that the induced diffeomorphism in $M \times \mathbb{R}$ is C^∞. We assume φ_μ to have a saddle point p_μ with stable and unstable separatrices $W^s(p_\mu)$ and $W^u(p_\mu)$, such that for $\mu = 0$ they have a nondegenerate tangency which unfolds generically with μ. Also, we assume that for μ near zero,

$\left|\det(d\varphi_\mu)_{p_\mu}\right| \neq 1$ and φ_μ is C^2-linearizable near p_μ. We note that these conditions are generic in a strong sense: among the one-parameter families φ_μ with homoclinic tangency for $\mu = 0$, there is an open and dense subset (with respect to the C^∞ topology) consisting of one-parameter families for which the above assumptions hold.

THEOREM 1. *For one-parameter families φ_μ, satisfying the above assumptions, there are nontrivial intervals $I \subset \mathbb{R}$, arbitrarily near zero, such that for generic $\mu \in I$, φ_μ has infinitely many hyperbolic periodic sinks (or sources).*

The proof of this theorem will occupy the remaining part of this appendix. First, we observe that our assumptions correspond exactly to the type of one-parameter family constructed in the second perturbation in the proof of the main result of Chapter 6; see Section 5. For the present theorem we need to show that the last perturbation (making $W^u(Q_2)$ and $W^s(P)$ transverse) is not necessary. We do this by returning to the Hénon-like diffeomorphisms; see Section 3 of the same chapter. In the second proposition of that section, we concluded, among other things, that $W^u(Q_m(\mu))$ crosses $W^s(P(\mu))$. All we have to show now is that these curves not only are crossing but even have transverse intersections. For this we need a result about stable manifolds of endomorphisms, in particular the continuous dependence (in the C^k topology) of these manifolds on the defining map. As mentioned in Appendix 1, a local version of such a result follows directly from the construction of local stable manifolds, if we use Perron's method which does not require taking inverse maps; see [**PM**,1982] or Prop. II.5 in [**S**,1978]. It can be formulated as follows in arbitrary dimension.

THEOREM 2 (LOCAL STABLE MANIFOLD). *Let $\varphi \colon \mathbb{R}^m \to \mathbb{R}^m$ be a C^k map, $k \geq 1$, with a hyperbolic fixed point $p \in \mathbb{R}^m$, i.e. no eigenvalue of $(d\varphi)_p$ has norm 1. For $\varepsilon > 0$ sufficiently small, the local stable manifold*

$$W^s_\varepsilon(p) = \{x \in \mathbb{R}^m \,|\, \left\|\varphi^i(x) - \varphi^i(p)\right\| \leq \varepsilon \quad \text{for all } i \geq 0, \quad \lim_{i \to \infty} \varphi^i(x) = p\}$$

is a C^k submanifold. Also there is a neighbourhood $\mathcal{U}$ of φ in the C^k topology and a continuous map $p \colon \mathcal{U} \to \mathbb{R}^m$ with $p(\varphi) = p$ such that for $\varphi' \in \mathcal{U}$, $p(\varphi')$ is a hyperbolic fixed point of φ' and such that the local stable manifold $W^s_\varepsilon(p(\varphi'))$, as a C^k manifold, depends continuously on φ', with respect to the C^k topology.

For our purpose we need a somewhat more global version of the continuous dependence, in the C^k sense, of the stable "manifold" on the map. One problem is that, globally, the stable set of a hyperbolic fixed point of a map need not be a manifold. For this reason, we shall restrict to (compact sets of) *regular points*, to be defined below.

Let φ, p, and $W_\varepsilon^s(p)$ be as in the above theorem. As usual we define the stable set as

$$W^s(p) = \{x \in \mathbb{R}^m | \lim_{i \to \infty} \varphi^i(x) = p\}.$$

Let, now, $K \subset W^s(p)$ be any compact set. Then, for some n, $\varphi^n(K) \subset W_\varepsilon^s(p)$. Now we assume that for each $x \in K$,

$$Im(d\varphi^n)_x + T_{\varphi^n(x)}(W_\varepsilon^s(p)) = \mathbb{R}^m.$$

In this case we say that the points of K are *regular* stable points (or regular points of $W^s(p)$). Note that this property is independent of n and ε (as long as $\varphi^n(x) \in W_\varepsilon^s(p)$). We have

$$(\varphi^n)^{-1}(W_\varepsilon^s(p)) \subset W^s(p),$$

and the first set is, near points of K, locally a manifold (due to regularity and the implicit function theorem). So for each point of K we have a k-jet of a manifold contained in $W^s(p)$. The set of these jets, in the k-jet space, is denoted by K^k. Combining the local stable manifold theorem with the above definitions and the implicit function theorem, we obtain the following extension of the previous theorem.

PROPOSITION 1. *For φ as in the above theorem and $K \subset W^s(p)$ a compact set of regular points of $W^s(p)$ with corresponding set K^k of k-jets, there are, on a possibly smaller neighbourhood $\mathcal{U}'$ of φ, continuous maps which assign to each $\varphi' \in \mathcal{U}'$ a hyperbolic fixed point $p(\varphi')$, a compact set $K(\varphi') \subset W^s(p(\varphi'))$ of regular points of $W^s(p(\varphi'))$ with corresponding set $K^k(\varphi')$ of k-jets. The maps $\varphi' \mapsto K(\varphi')$ and $\varphi' \mapsto K^k(\varphi')$ are continuous in the sense that there are homeomorphisms $h(\varphi')$, $h^k(\varphi')$, depending continuously on φ', mapping K, respectively K^k, to $K(\varphi')$, respectively $K^k(\varphi')$.*

We apply the above result to the Hénon-like diffeomorphisms (see Chapter 6, Section 3), and especially to those which are near the endomorphism $\varphi_{-2,0}(x,y) = (y, y^2 - 2)$ (see Figure A4.1). This endomorphism has a fixed point $p = (2,2)$. The stable set $W^s(p)$ of this fixed point contains the lines $\{y = \pm 2\}$. In fact these lines contain the only regular points of $W^s(p)$: the next line $\{y = 0\}$ of $W^s(p)$ consists of nonregular points.

This means that for μ_1, μ_2 near $(-2,0)$, $\mu_2 > 0$, the diffeomorphism φ_{μ_1,μ_2}

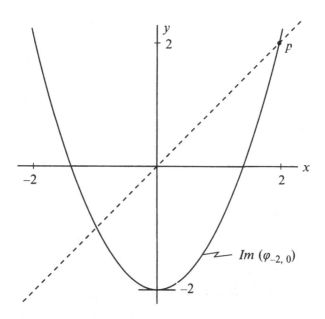

Figure A4.1

(see Chapter 6, Section 3) has a saddle point $P(\mu)$ near P whose stable manifold contains arcs which are, by the above result, C^2-close to the segments $\{(x,y)|\,|x| \le 3,\ y = \pm 2\}$. The fact that the arcs are C^2-close means that the curvature is close to zero.

Returning to Proposition 2 of Chapter 6, Section 3, we have for $\mu = (\mu_1,\mu_2) \in \gamma$ and close to $(-2,0)$ the situation indicated in Figure A4.2.

We also indicate the periodic point $Q_m(\mu)$ which was introduced in the above mentioned proposition in Chapter 6. In the small neighbourhood of $(0,-2)$ (which can be taken smaller as $\mu \in \gamma$ comes closer to $(-2,0)$) we have the situation (tangency of $W^u(P(\mu))$ and $W^s(P(\mu))$ and two crossings of $W^u(Q_m(\mu))$ and $W^s(P(\mu))$) as indicated. Also the curvature of the branch of $W^u(Q_m(\mu))$ near $(0,-2)$ is close to the curvature of Im $(\varphi_{-2,0})$ at $(0,-2)$ which is 2. This means that $W^u(Q_m(\mu))$ and $W^s(P(\mu))$ can only cross with two *transverse intersections*.

Now the proof of our main theorem in this appendix is complete. □

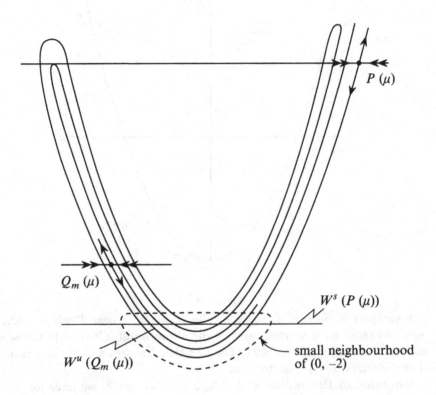

Figure A4.2

Hyperbolicity and the creation of homoclinic orbits

By J. Palis and F. Takens*

Abstract

We consider one-parameter families $\{\varphi_\mu; \ \mu \in \mathbf{R}\}$ of diffeomorphisms on surfaces which display a *homoclinic tangency* for $\mu = 0$ and are *hyperbolic* for $\mu < 0$ (i.e., φ_μ has a hyperbolic nonwandering set); the tangency unfolds into *transversal homoclinic orbits* for μ positive. For many of these families, we prove that φ_μ is also hyperbolic for most small positive values of μ (which implies much regularity of the dynamical structure). A main assumption concerns the *limit capacities* of the basic set corresponding to the homoclinic tangency.

1. Introduction

In 1899, Poincaré first detected the existence of what he called *homoclinic orbits* (doubly asymptotic solutions) when studying the restricted three-body problem [12]. He expressed amazement about the dynamical complexity implied by the presence of what seemed to be a rather simple phenomenon. To recall this concept, let M be a 2-manifold, $\varphi: M \to M$ a diffeomorphism and p a fixed point for φ. We say that x is a *homoclinic point* associated with p if $\varphi^n(x) \to p$ for both $n \to +\infty$ and $n \to -\infty$; that is, when x has p as its ω and α-limit sets. When p is a periodic point of period k, we can define homoclinic points or orbits in a similar way by considering φ^k. We will be particularly interested in the case where p is a fixed (hyperbolic) *saddle*, i.e., when one eigenvalue of $d\varphi(p)$ has norm bigger than one and the other smaller than one. A homoclinic point means, in this situation, a point of intersection of the *stable* and *unstable manifolds* of p (see [3]). Depending on this intersection, we have a *transversal homoclinic point* or a *homoclinic point of tangency*. In both cases it is easy to see that a homoclinic point, say x, is *nonwandering*: for any given neighborhood U of x, there is an arbitrarily large integer n such that $\varphi^n U \cap U \neq \varnothing$. The

---

*The first author acknowledges hospitality and financial help from Rijksuniversiteit Groningen/Z.W.O./K.N.A.W. during the preparation of this paper. Similarly, the second author is thankful to IMPA/CNPq.

union of the nonwandering points is called the nonwandering set, which is closed and denoted by $\Omega(\varphi)$.

Around 1935 Birkhoff proved the following relevant result in this context: every transversal homoclinic orbit of a 2-dimensional diffeomorphism is accumulated by periodic orbits [1]. In 1965, Smale generalized this result both in two and higher dimensions by showing that a transversal homoclinic orbit is contained in a hyperbolic set (in this case a "horseshoe") in which the periodic orbits are dense [15]. By a *hyperbolic set* we mean a closed, φ-invariant subset Λ of M such that the tangent bundle $T_\Lambda M$ splits into $d\varphi$-invariant sub-bundles E^s and E^u on which $d\varphi$ is uniformly contracting and uniformly expanding, respectively. A key fact is that through each point $x \in \Lambda$ we can construct a pair of "leaves", which are called stable and unstable manifolds of x and denoted by $W^s(x)$ and $W^u(x)$. The union of these leaves forms (partially defined) *foliations* denoted by $\mathscr{F}^s(\Lambda)$ and $\mathscr{F}^u(\Lambda)$ (see [3]). They have smooth leaves (C^r if φ is C^r), they are φ-invariant (φ sends leaves into leaves) and they are characterized as follows: given $x \in \Lambda$, the leaf $W^s(x)$ or $\mathscr{F}^s(x)$ consists of points y such that $d(f^n, y, f^n x) \to 0$ as $n \to \infty$, where d is a distance function of M; dually for $\mathscr{F}^u(\Lambda)$.

The results of Birkhoff and Smale confirm Poincaré's observation about how rich the dynamics of a transformation seemed to be because of the mere existence of a (transversal) homoclinic orbit. In the present paper we want to analyse, for a family of C^2 diffeomorphisms $\{\varphi_\mu; \mu \in \mathbf{R}\}$ of a closed surface, the global dynamical implications of creating a homoclinic orbit (of tangency) say, for $\mu = 0$ and letting it unfold to become transversal. It is well-known that such a process implies infinitely many changes (bifurcations) of the dynamical structure no matter how small we take the interval $(-\delta, \delta)$ of a parameter range. The set of values of μ for which these changes occur (*bifurcation set*) even contains intervals! This important fact is due to Newhouse [8]. However, as we show here, in many relevant cases the *relative measure* of the bifurcation set is very small, asymptotically going to zero with the diameter of the parameter range. Moreover φ_μ has no "new" attractors when μ is not in the bifurcation set, so not much "chaos" occurs in this 1-parameter family for μ near zero. We now turn to a description of the one-parameter families we intend to study. First, we assume that $\Omega(\varphi_\mu)$ is *persistently hyperbolic* for $\mu < 0$. (We will see in Chapter 4 that this implies Ω-stability.) Also, we suppose that $\Omega(\varphi_0)$ is a (closed) hyperbolic set together with a newly created homoclinic orbit of tangency. A second main and perhaps crucial ingredient concerns the *limit capacities* (capacity dimensions) of the stable and unstable foliations of the (basic) hyperbolic set that gives rise to the homoclinic tangency; their sum should be smaller than one.

Let us recall a few relevant concepts including the ones we have just mentioned, and then present the precise statement of our main result. For a diffeomorphism $\varphi: M \to M$, a closed subset $\Lambda \subset \Omega(\varphi)$ is called a *basic set* if it is hyperbolic, transitive (it has a dense orbit) and its subset of periodic orbits is dense. The basic sets that appear in our context are open in $\Omega(\varphi_\mu)$ for $\mu \leq 0$, except that for $\mu = 0$ there is *one* which is a limit as $\mu \nearrow 0$ of such a basic set for $\mu < 0$ and which has a saddle point with a homoclinic tangency. See the first section of Chapter 4 for more details. We define φ to be *hyperbolic* if $\Omega(\varphi)$ is hyperbolic, and *persistently hyperbolic* if every $\tilde{\varphi}$ which is C^r near φ ($r \geq 1$) is also hyperbolic. Here we will consider C^2 diffeomorphisms of a closed surface M endowed with the C^2 topology. We will show in Chapter 4 that if φ is persistently hyperbolic then φ is Ω-stable. That is, for every $\tilde{\varphi}$ near φ there is a homeomorphism $h: \Omega(\varphi) \to \Omega(\tilde{\varphi})$ such that $h\varphi = \tilde{\varphi}h$. A relevant fact about a C^2 diffeomorphism on a surface concerns the stable and unstable foliations of a basic set Λ: $\mathscr{F}^s(\Lambda)$ and $\mathscr{F}^u(\Lambda)$ can be extended to C^1 invariant foliations defined on a full neighborhood of Λ ([3]). This fact allows us to define intrinsically the following concept of "transverse dimensions" for $\mathscr{F}^s(\Lambda)$ and for $\mathscr{F}^u(\Lambda)$, which we call stable and unstable limit capacities of Λ. First, for a compact metric space C we define the *limit capacity* $d(C)$ as $\limsup \mathrm{Log}\, n(C, \varepsilon)/\mathrm{Log}\, 1/\varepsilon$ as $\varepsilon \to 0$, where $n(C, \varepsilon)$ is the minimum number of balls of radius ε needed to cover C. Notice that $d(C) = d(f(C))$ if f is a Lipschitz homeomorphism with Lipschitz inverse. For $\mathscr{F}^s(\Lambda)$ we consider, through a point of a leaf, a small transversal segment L and let C_L be the set of intersection points of the leaves with L. We then define the *stable limit capacity* $d^s(\Lambda)$ as $d(C_L)$. The above remark, on extending $\mathscr{F}^s(\Lambda)$ to a C^1 foliation on a neighborhood of Λ, shows as in [8], [11] that this definition does not depend on the transversal segment L. A similar procedure for $\mathscr{F}^u(\Lambda)$ gives the *unstable limit capacity* $d^u(\Lambda)$. If Λ is of *saddle type* (i.e., neither the leaves of $\mathscr{F}^s(\Lambda)$ nor those of $\mathscr{F}^u(\Lambda)$ are contained in Λ) and does not consist of a single periodic orbit, then C_L is a Cantor set and thus the stable limit capacity $d^s(\Lambda)$ can take any value between zero and one. Similarly for $d^u(\Lambda)$.

We now describe the parametrized families of diffeomorphisms which we consider. We say that a 1-parameter family $\varphi_\mu: M \to M$ of C^2 diffeomorphisms of a closed surface M has a *homoclinic Ω-explosion* at $\mu = 0$ if:

i) For $\mu < 0$, φ_μ is persistently hyperbolic;

ii) For $\mu = 0$, the nonwandering set $\Omega(\varphi_0)$ consists of a (closed) hyperbolic set $\tilde{\Omega}(\varphi_0) = \lim_{\mu \nearrow 0} \Omega(\varphi_\mu)$ together with a homoclinic orbit of tangency $\mathcal{O}$ associated with a fixed saddle point p, so that $\Omega(\varphi_0) = \tilde{\Omega}(\varphi_0) \cup \mathcal{O}$; the product of the eigenvalues of $d\varphi_0$ at p is different from one in norm;

iii) The separatrices have quadratic tangency along $\mathcal{O}$ unfolding generically (see Chapter 2); $\mathcal{O}$ is the only orbit of tangency between stable and unstable separatrices of periodic orbits of φ_0.

Since $\Omega_\mu = \Omega(\varphi_\mu)$ is hyperbolic for $\mu < 0$, we can write Ω_μ as a finite disjoint union of basic sets $\Lambda_1, \ldots, \Lambda_r$ (see Section 1, Chapter 4). From (ii) above, we have $p = \lim_{\mu \nearrow 0} p(\mu)$ where $p(\mu) \in \Lambda_i(\mu)$ for some basic set $\Lambda_i(\mu)$. Let us denote $\Lambda_i(0) = \lim_{\mu \nearrow 0} \Lambda_i(\mu)$ simply by Λ and call it the basic set associated with the homoclinic tangency $\mathcal{O}$. Finally, let us denote by B the set of values of $\mu > 0$ for which φ_μ is not persistently hyperbolic. As mentioned before, we will see in Chapter 4 that this set contains the set of μ's for which φ_μ is not Ω-stable (Ω-bifurcation set).

MAIN THEOREM. *Let* $\{\varphi_\mu;\ \mu \in \mathbf{R}\}$ *be a family of diffeomorphisms of M with a homoclinic Ω-explosion at $\mu = 0$. Suppose that $d^s(\Lambda) + d^u(\Lambda) < 1$, where Λ is the basic set of φ_0 associated with the homoclinic tangency. Then*

$$\lim_{\delta \to 0} \frac{m(B \cap [0, \delta])}{\delta} = 0$$

where m denotes Lebesgue measure.

The theorem above considerably extends a result in [7] in two ways. First, in that paper the families $\{\varphi_\mu\}$ were such that $\Omega(\varphi_\mu)$ was finite (and hyperbolic), which we do not assume here. To deal with a non-trivial basic set (one that is not reduced to a periodic orbit) giving rise to a homoclinic tangency, we introduce the (transverse) limit capacities of the stable and unstable foliations. We then draw the conclusion that, if these capacities are not too large, we get mainly hyperbolic diffeomorphisms right after the unfolding of the tangency. Second, the result in [7] stated that $\liminf_{\delta \to 0} m(B \cap [0, \delta])/\delta = 0$, instead of the full limit as in the present paper. Another related, but more recent, work is [11], where we dealt with similar questions in the context of heteroclinic tangencies and larger cycles, i.e., when more than one basic set is involved (see Chapter 4). Also in that last paper we only proved that $\liminf_{\delta \to 0} m(B \cap [0, \delta])/\delta = 0$.

The content of the next chapters of this paper are as follows. In Chapter 2 we present a catalog of all possible types of 1-parameter families with homoclinic Ω-explosion that can exist on surfaces. The proof of the main result is in Chapter 4. It involves determining the locus of the nonwandering set and then establishing its hyperbolicity when there is a certain (positive) minimal angle between the leaves of the stable and unstable foliations of the basic set involved. The fact that this condition is frequently satisfied for μ positive and small depends heavily on Chapter 3. There, the assumption on the capacities is used to analyse intersection of (Cantor) sets when shifting one across the other. A special difficulty in our context is that the sets depend (although differentiably) on the parameter μ.

We finish by conjecturing a converse to our main result implying a strong correlation between hyperbolicity and limit capacity.

Conjecture. Suppose $\{\varphi_\mu\}$ has a homoclinic Ω-explosion at $\mu = 0$ and let Λ be the basic set associated with the homoclinic tangency of φ_0. Let $d^s(\Lambda)$ and $d^u(\Lambda)$ be the stable and unstable limit capacities of Λ. If $d^s(\Lambda) + d^u(\Lambda) > 1$, then

$$\liminf_{\delta \to 0} \frac{m(B \cap [0, \delta])}{\delta} > 0.$$

As usual, one may need some "generic" conditions on the family $\{\varphi_\mu\}$ to be able to verify the conjecture.

We can construct examples that corroborate this conjecture. For this we assume the sum of the stable and unstable *Hausdorff dimensions* to be greater than one; however, in a personal communication, F. Ledrapier and L. S. Young told us that in our context, limit capacity and Hausdorff dimension are equal. To get the examples we made much use of results in [4] about sets of fractional dimensions.

2. Classification of homoclinic Ω-explosions in dimension 2

1. *Introduction.* We consider one parameter families $\varphi_\mu\colon M \to M$ of diffeomorphisms with homoclinic Ω-explosion at $\mu = 0$ as defined in the introduction. We also denote the saddle point, corresponding to the homoclinic tangency for φ_0, by p; the continuation of this saddle point for φ_μ is denoted by p_μ.

We recall that the homoclinic tangency is supposed to satisfy some generic properties. In the first place the *tangency* should be of *first order* not of *second order*. This means that if r is in the orbit of tangency, then $W^u(p)$ and $W^s(p)$ have different curvature in r. We also assume that this *tangency unfolds generically*; e.g. see [9]. This means that near a point of tangency of φ, the separation between $W^u(p_\mu)$ and $W^s(p_\mu)$ is as indicated below:

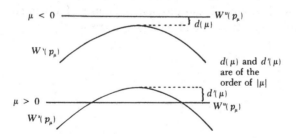

Finally, we assume that the product of the eigenvalues of $(d\varphi_0)_p$ has norm different from 1.

In the present chapter we shall distinguish these homoclinic Ω-explosions according to the following criteria, all of which refer to φ_0 (from now on we also denote φ_0 by φ):

a) *The signs of the eigenvalues* of $(d\varphi)_p$: here we have three possibilities, $+ +$, $+ -$, and $- -$ (in the $+ -$ case one can of course distinguish between the cases where the positive eigenvalue is expanding or contracting but one can change from one case to the other by replacing φ by φ^{-1});

b) *The sides of tangencies*: by this we mean the following: if $q \in W^s_{\text{loc}}(p)$ and $r \in W^u_{\text{loc}}(p)$ are in the orbit of tangency (loc indicates that we take only a small interval around p in the corresponding separatrix), then the position of $W^u(p)$ near q and $W^s(p)$ near r can be as in the figures below:

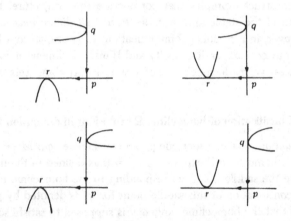

We refer to these differences as the different sides of tangency (if at least one of the eigenvalues of $(d\varphi)_p$ is negative we should take two successive points in the orbit of tangency in both $W^u_{\text{loc}}(p)$ and $W^s_{\text{loc}}(p)$));

c) *The connections*: these concern the distinction that still can be made if the sides of tangencies are known; for example if they are as in the first of the above cases, then the (global) unstable separatrix of p can still connect in two different ways with the piece of stable separatrix near q:

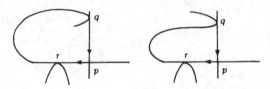

In these cases we speak of different connections (each branch of a stable or unstable separatrix has an orientation, given by the dynamics; points move in the positive direction; the different connections can be distinguished according to whether these two orientations agree or disagree in points of the orbit of tangency).

In all cases which may exist according to the above subdivision, we want to determine whether they can occur as a homoclinic Ω-explosion, and if so:
   – if the ambient manifold can be orientable;
   – if the saddle point p can be part of a non-trivial basic set or not.
In some cases the answer to these questions depends also on the so-called eigenvalue conditions. If λ, σ are the eigenvalues of $(d\varphi)_p$, with $0 < |\lambda| < 1 < |\sigma|$, we say that the *contracting*, resp. *expanding*, *eigenvalue* is *dominating* if $|\Lambda| \cdot |\sigma| < 1$, resp. $|\Lambda| \cdot |\sigma| > 1$. As we mentioned before we exclude the case $|\lambda| \cdot |\sigma| = 1$ (although this is not always necessary).

The rest of this chapter is organized as follows. In Section 2 we consider a number of typical cases and establish the answers to the above questions. These typical cases were selected in such a way that all the different arguments are displayed. Then, in Section 3 we treat all cases (according to the above subdivision), referring back to Section 2 for the main arguments. In Section 4 we construct examples of all the remaining cases which can occur as homoclinic Ω-explosions.

2. *Typical cases.*

   A. *Premature tangencies.* We shall show that the case with eigenvalues $+ +$, and sides as in the figure below

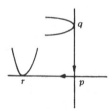

cannot occur as homoclinic Ω-explosion because then φ_μ would have orbits of tangency of $W^u(p_\mu)$ and $W^s(p_\mu)$ for negative values of μ arbitrarily close to zero.

In order to prove this we take as before $q \in W^s(p)$ and $r \in W^u(p)$ both in the orbit of tangency of φ. We also take $\tilde{q} = \varphi^n(q)$ for some (big) $n > 0$. We consider, for some neighborhood U of p the components of $W^u(p) \cap U$ and $W^s(p) \cap U$ containing r, q, and $\tilde{q}$; see the figure below on the left.

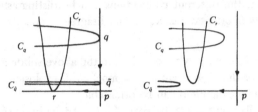

The components of $W^u(p) \cap U$ and $W^s(p) \cap U$, of r, q, and $\tilde{q}$, not containing p, are denoted by C_r, C_q, and $C_{\tilde{q}}$. For small negative values of μ one can follow these components and one gets, for example, a situation as in the above figure on the right: C_r and C_q moved slightly but $C_{\tilde{q}}$ almost disappeared. In fact if one lets μ go down from zero to negative values, $C_{\tilde{q}}$ moves to the left; the bigger n is in $\tilde{q} = \varphi^n(q)$, the faster $C_{\tilde{q}}$ moves. For $\mu = 0$, C_r and $C_{\tilde{q}}$ have four points of intersection. When they disappear there is a value of μ where they are tangent. Since $C_r \subset W^s(p_\mu)$ and $C_{\tilde{q}} \subset W^u(p_\mu)$, there is a sequence of negative values μ_i converging to zero for which $W^u(p_{\mu_i})$ and $W^s(p_{\mu_i})$ have an orbit of tangency and this is incompatible with φ_{μ_i} being hyperbolic. This argument was provided to us by J. Gheiner.

    B. *Premature creation of horseshoes.* We consider the case where both eigenvalues of $(d\varphi)_p$ are positive and where the sides of tangencies and the connections are as indicated below (λ and σ are the contracting and expanding eigenvalues of $(d\varphi)_p$).

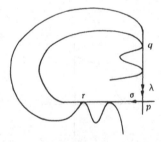

    We claim that this case cannot occur as a homoclinic Ω-explosion if $\lambda^{-1} > \sigma$, i.e. if the contracting eigenvalue is dominating. In order to prove this we show that, if $\lambda^{-1} > \sigma$, $\varphi = \varphi_0$ has horseshoes which disappear when moving μ to small negative values. Of course this implies that there are also premature tangencies (as, also, premature tangencies imply premature creations or destructions of horseshoes). We only distinguish these cases by names, indicating the most easily visible reason why they cannot occur as homoclinic Ω-explosion.

We take C^1-linearizing coordinates in a neighborhood of p; for the existence of such coordinates see [2]. Let $r \in W_{\text{loc}}^u(p)$ be a point in the orbit of tangency; without loss of generality we may assume that r has coordinates $(-1, 0)$ (in the linearizing coordinates). Let R be a rectangle as indicated below: one side is part of $W_{\text{loc}}^u(p)$ and contains r in its interior; the length may be chosen for example equal to $1/2$; the height d is small. We consider negative iteratates of R. For $i < \text{Log } d/\text{Log } \lambda$, the height of $\varphi^{-1}(R)$ is smaller than one; let $i_0(d)$ be the

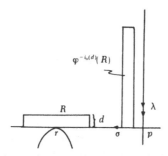

smallest integer for which this inequality is not satisfied. Then the distance from $\varphi^{-i_0(d)}(R)$ to $W_{\text{loc}}^s(p)$ is

$$\approx \sigma^{-i_0(d)} \approx d^{-\log \sigma/\log \lambda}$$

($\approx$ means: equal up to a multiplicative factor which is bounded and bounded away from zero, uniformly in d). Since $0 < \lambda < 1 < \sigma$ and $\lambda^{-1} > \sigma$, the distance from $\varphi^{-i_0(d)}(R)$ to $W_{\text{loc}}^s(p)$ is much bigger than d (for d small). For some integer j_0, independent of d, $r \in \varphi^{-j_0}(W_{\text{loc}}^s(p))$; here $W_{\text{loc}}^s(p)$ is the interval in $W^s(p)$ consisting of points at distance less than one from p. Then the distance from $\varphi^{-i(d_0)-j_0}(R)$ to $\varphi^{-j_0}(W_{\text{loc}}^s(p))$ is also much bigger than d (for d small). So we have the following situation

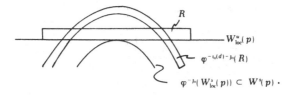

This means that in R we have a horseshoe, or at least a nonempty set of nonwandering points $\Lambda = \bigcap_{n \in \mathbb{Z}} \varphi^{n \cdot (i_0(d)+j_0)}(R)$. For $\mu < 0$ we define $\Lambda_\mu = \bigcap_{n \in \mathbb{Z}} \varphi^{n \cdot (i_0(d)+j_0)}(R)$ (we assume that $W_{\text{loc}}^u(p_\mu)$ is independent of μ, so that we can choose R also independent of μ). We have $\Lambda_0 = \Lambda \neq \varnothing$; for some $\mu_0 < 0$, $\Lambda_{\mu_0} = \varnothing$. Also, for $\mu \in (\mu_0, 0)$, $\Lambda_\mu \cap \partial R = \varnothing$. From this it is clear that there are Ω-bifurcations between μ_0 and 0. Since $|\mu_0|$ is of the order $d^{-\text{Log }\sigma/\text{Log }\lambda}$,

$|\mu_0|$ can be made arbitrarily small by taking d small. This proves that φ_μ has Ω-bifurcations for negative values of μ arbitrarily close to zero (when we apply the above argument for values of d going to zero).

     C. *Infinitely many circles.* We consider the case where both eigenvalues of $(d\varphi)_p$ are positive and where the sides of tangencies and connections are as indicated below.

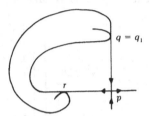

We assume that for $\mu \leq 0$, p_μ is not part of a non-trivial basic set. Then this situation cannot occur in a homoclinic Ω-explosion on a closed surface.

    We prove this by contradiction. Let γ^s, γ^u be the branches of $W^s(p)$, $W^u(p)$ containing the orbit of tangency. Since for $\mu \leq 0$, p_μ is not part of a non-trivial basic set, γ^s and γ^u only meet in p and in the orbit of tangency. As indicated, q_1 is a point of the orbit of tangency and $q_i = \varphi^{i-1}(q_1)$. We define the closed curve C_i as the union of the segments in γ^u and γ^s joining q_{2i-1} and q_{2i}. Since γ^u and γ^s only meet in p and in the orbit of tangency, the closed curves C_i are double-point free and pairwise disjoint. We then prove that the complement of $\cup_{i=1}^k C_i$ is connected for any k (assuming that φ is defined on a connected surface). If this is proven we are done: for each closed surface M there is a number $n(M)$ such that any collection of more than $n(M)$ closed curves in M separates M in at least two components; e.g. see [13].

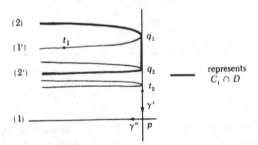

represents
$C_1 \cap D$

In the above figure we show part of $W^u(p)$ and $W^s(p)$ in a neighborhood D of p which is homeomorphic to a disc. Outside D the components of $\gamma^u \cap D$ are connected as indicated: $(1) \to (1')$, $(2) \to (2')$, etc. If the complement of C_1

were not connected, then the points t_1 and t_2 (see the above figure) would be in different components. But the segments in γ^u and γ^s joining t_1 with p and p with t_2 are disjoint from C_1; so the complement of C_1 is connected.

The arc in γ^u from t_1 to p is disjoint from $C_2, \ldots, C_k$. The arc in γ^s from p to t_2 becomes disjoint from $C_2, \ldots, C_k$ if we shift it slightly to the right. This means that the complement of $U_{i=1}^k C_i$ has just as many components as the complement of $U_{i=2}^k C_i$. Now the connectedness of the complement of $U_{i=1}^k C_i$ follows by induction. Note that we only used the fact that $\gamma^u \cap \gamma^s$ is the union of p and the orbit of tangency $\mathcal{O}$. This is weaker than the assumption that for $\mu \leq 0$, p_μ is not part of a non-trivial basic set.

D. *Basic sets and negative eigenvalues.* If a saddle point p is part of a non-trivial basic set Λ, then we indicate the (local) position of Λ with respect to p by the quadrant(s) (or sector(s)), separated by $W^u_{loc}(p)$ and $W^s_{loc}(p)$, in which there are points of $\Lambda \cap U$ for any small neighborhood U of p. In a situation where all homoclinic points are transverse, as for φ_μ, $\mu < 0$, we have the following. If γ^u and γ^s are branches of $W^u(p)$ and $W^s(p)$, then $\gamma^u \cap \gamma^p \supsetneq \{p\}$ if and only if there is (part of) a basic set in the quadrant bounded by γ^u and γ^s. Also, if p is part of a non-trivial basic set Λ then $W^u(p)$ and $W^s(p)$ are accumulating on themselves at least from one side as indicated in the following figure.

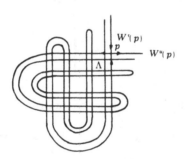

Returning to our 1-parameter families φ_μ with homoclinic Ω-explosion for $\mu = 0$ and φ_μ hyperbolic for $\mu < 0$, we see that if the sides of tangencies are as indicated below

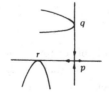

so that for $\mu < 0$ we have

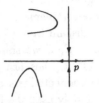

then the only quadrant where there might be a basic set is the upper right quadrant; for any other quadrant there would be tangencies of $W^u(p_\mu)$ and $W^s(p_\mu)$ for negative values of μ arbitrarily close to zero. For this reason, once the sides of tangencies are given, there is at most one quadrant which could be occupied by a basic set.

If at least one of the eigenvalues of $(d\varphi)_p$ is negative, then p cannot be part of a non-trivial basic set, because such a basic set would occupy at least two quadrants.

E. *Orientability.* We consider, as an example, a 1-parameter family φ_μ with homoclinic Ω-explosions, such that both eigenvalues of $(d\varphi)_p$ are positive and such that the sides of tangencies are as indicated below.

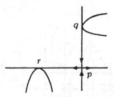

In this case we can still make the connections in two ways:

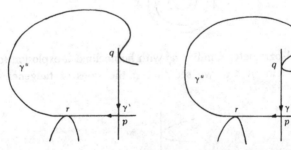

As observed in the introduction of this chapter, one can distinguish these cases by orienting the branches γ^u and γ^s in such a way that φ moves the points in the positive direction. These orientations are intrinsically defined in the sense that they are preserved by conjugacies. With these orientations of γ^u and γ^s one easily sees the differences between the above two cases: at a point of tangency, say q, the two orientations of γ^u and γ^s are in the same direction in the first case and in opposite directions in the second case. Clearly the same must hold at other points of tangency like r. In this way we know the orientation of the piece of γ^s near r.

There is another way of seeing the difference between the two cases: if we take a diffeomorphism from a neighborhood of r to a neighborhood of q, respecting γ^u and γ^s, with orientations, then this diffeomorphism is orientation reversing in the first case and orientation preserving in the second case. From this it follows that the first case cannot occur if the ambient surface is orientable: φ is orientation preserving since $(d\varphi)_p$ has positive eigenvalues and for some positive i, φ^i maps r to q preserving γ^u and γ^s with orientations.

In case $(d\varphi)_p$ has one negative eigenvalue (and hence φ cannot be orientation preserving) one can apply a similar argument using φ^2 instead of φ.

3. *General classification.* In this section we go systematically through all the cases as distinguished by: signs of eigenvalues, sides of tangencies, connections, and where necessary eigenvalue condition.

I. *Eigenvalues* $++$. In this case there are four possibilities of sides of tangencies:

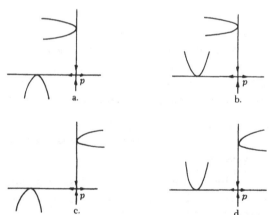

Before distinguishing according to the different connections, we observe that (b) cannot occur due to premature tangencies and that the cases (a) and (d) can be interchanged by replacing φ by φ^{-1}. So we only have to consider the cases (a) and (c). Consideration of the different connections again gives four cases.

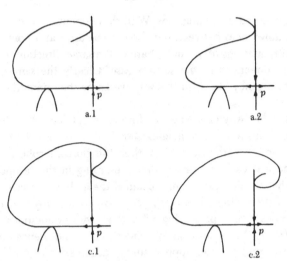

Case (a.1) can only occur if the expanding eigenvalue dominates the contracting eigenvalue; see (B) on the "premature creation of horseshoes"; also, it was given as an example in [5]. This case can occur both when p is part of a non-trivial basic set and when p is an isolated point in $\tilde{\Omega}(\varphi)$ (the hyperbolic part of the non-wandering set of φ). The ambient surface can be orientable or non-orientable in this case.

Case (a.2) Here p is part of a non-trivial basic set; see (C) *infinitely many circles*. There must be a basic set in the sector between γ^u and γ^s, the branches of $W^u(p)$ and $W^s(p)$ containing the orbit of tangency (because these branches must also have transverse intersections). This sector is however not allowed according to *basic sets and negative eigenvalues*. So this case cannot occur.

Case (c.1) was already considered in [5] as an example in the projective plane. It cannot occur in an orientable surface.* The saddle point p may be both isolated or not in $\tilde{\Omega}(\varphi)$.

Case (c.2) Again by *infinitely many circles* this case can only occur when p is part of a non-trivial basic set. As such it occurs in both orientable and non-orientable surfaces.

II. *Eigenvalues* $+ -$. In this case there are two possibilities of sides of tangencies. We take the expanding eigenvalue as negative; the other case can then be obtained by replacement of φ by φ^{-1}.

---

*See *orientability*.

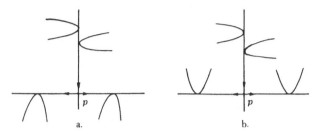

Case (b) cannot occur because of premature tangencies. In case (a) we can make the connections in four different ways.

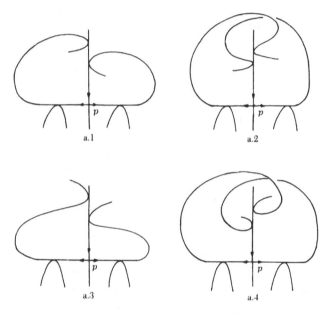

Case (a.1) can only occur if the expanding eigenvalue is dominating (see *premature creation of horseshoes*) and if p is isolated in $\tilde{\Omega}(\varphi)$ (see *basic sets and negative eigenvalues*). This case can occur in both orientable and non-orientable surfaces.

Case (a.2) can only occur on non-orientable surfaces when p is isolated in $\tilde{\Omega}(\varphi)$.

Case (a.3) *and* (a.4) cannot occur as homoclinic Ω-explosion. This follows from the argument in "infinitely many circles" together with the fact that p cannot be part of a basic set.

III. *Eigenvalues* $--$. In this case there are two possibilities of sides of tangencies.

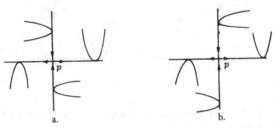

a.   b.

Case (b) cannot occur because of premature tangencies. In case (a) there are four different ways to make connections.

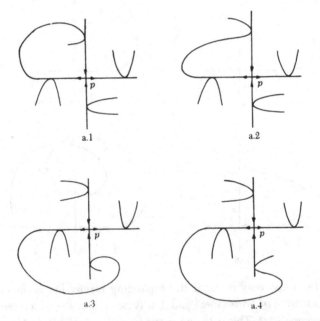

a.1   a.2

a.3   a.4

When φ is replaced by φ^{-1}, the cases (a.1) and (a.2) change to the cases (a.4) and (a.3) respectively. Both cases (a.1) and (a.2) (and hence also (a.3) and (a.4)) can be excluded because of premature creation of horseshoes. The argument goes as follows.

If the contracting eigenvalue dominates, we can immediately apply the arguments from the previous section. So we assume that the expanding eigenvalue dominates. We take a rectangle R, as indicated below, with small thickness d.

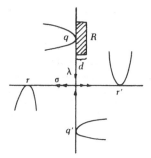

We assume that the point q of tangency is at distance one from p in C^1-lineariz-ing coordinates. For i, such $\varphi^{-2i}(q) = r$, $R' = \varphi^{-2i}(R)$ has the following shape:

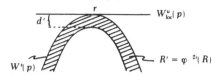

with d' of the same order as d. We denote the contracting and ex-panding eigenvalues of $(d\varphi)_p$ by λ and σ. Applying φ^{-2j} to R', where $j \approx \mathrm{Log}(2d')/2\,\mathrm{Log}\,|\lambda|$ we obtain $R'' = \varphi^{-2j}(R')$ as in the figure below.

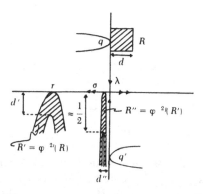

The maximal distance from R'' to $W^s_{\mathrm{loc}}(p)$ is d''. Due to the eigenvalue condition, d'' is much smaller than d for d sufficiently small. From this it follows that $\varphi^{-1}(R'') = \varphi^{-2j-2i-1}(R)$ intersects R in such a way that there has to be a horseshoe, which disappears if μ moves to negative values; see "premature creation of horseshoes".

4. *List of possible homoclinic Ω-explosions.* Here we indicate examples of 1-parameter families of diffeomorphisms φ_μ with homoclinic Ω-explosion for all cases which were not excluded in Section 3. We shall do this by making in each case a sketch of the stable and unstable separatrices of the saddle point p of the diffeomorphism $\varphi = \varphi_0$. Since each case which can be realized on an orientable surface can also be realized on a non-orientable surface, we only give examples on non-orientable surfaces when necessary, i.e. when there is no "orientable example".

I. *Eigenvalues* $++$.

Case (a.1): p isolated in $\tilde{\Omega}(\varphi)$.

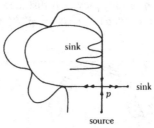

In order to see that this tangency is really a first bifurcation, we give a more careful description of the 1-parameter family φ_μ. We start with a diffeomorphism Φ as indicated below, with $\Omega(\Phi)$ consisting of one source and two sinks besides

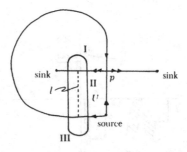

the saddle p. Let l be a line segment from $W^u(p)$ to $W^s(p)$, and U be a small neighbourhood of l as indicated. By $W^u(p)$ and $W^s(p)$, U is subdivided into three regions which are denoted by I, II, and III. We obtain our 1-parameter family by composing Φ with σ_μ, a 1-parameter family of diffeomorphisms with support in U which we describe below; $\varphi_\mu = \sigma_\mu \circ \Phi$.

For $\mu \leq -1$, σ_μ is the identity. For $\mu > -1$, σ_μ pushes the unstable separatrix down so that for $\mu = 0$ we have a generically unfolding tangency of $\sigma_\mu(W^u(p))$ and $W^s(p)$. Also we require that for any point x in region I, and

any $\mu \le 0$

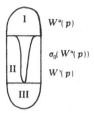

the distance of $\sigma_\mu(x)$ above $W^s(p)$ is at least equal to the distance of x above $W^u(p)$. We have to show that in this case φ_μ, $\mu < 0$, has no new non-wandering orbits. For this we only have to consider the orbits which continue to visit U: any orbit which does not pass through U, or which, after passing through U a finite number of times, never returns to U cannot be a new nonwandering orbit.

Observe that under forward iterations of Φ, the points of the regions I and III never return to U, while points of region II may return to region I. For $\mu < 0$, σ_μ maps region I inside the regions I and II, maps region II inside the regions II and III, and maps region III inside itself. From this it follows that the only possibility of having new nonwandering orbits is with points in region II which are mapped by Φ^n ($n > 0$) back into region I and then by σ_μ again to region II. It is however the eigenvalue condition which assures that such orbits are wandering, at least if U is small enough. So this proves that we have a first bifurcation at $\mu = 0$.

In the other six cases below the arguments are the same. In some cases, exactly in those cases where there is no eigenvalue condition, the argument simplifies somewhat. Then, just from the analysis of which transitions are possible to realize through positive iterations of Φ, and through σ_μ, it follows that no orbit of φ_μ, $\mu < 0$, continues to come back to U.

In the cases below we indicate the line l along which to carry out the deformation as well as the modified $W^u(p)$ near l; a main point is that l is disjoint from the positive iterates $\Phi^n(l)$.

Case (a.2); p part of a non-trivial basic set.

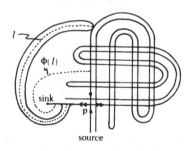

This is obtained from the usual horseshoe by a modification of $W^u(p)$ as indicated.

Case (c.1); p isolated in $\tilde{\Omega}(\varphi)$.

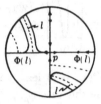

This is an example in the projective plane; in the boundary of this disc, antipodal points are identified.

Case (c.2); p part of a non-trivial basic set.

This case is more complicated in the sense that we have to make use of a "type-3 horseshoe", i.e., a horseshoe obtained by mapping a square D over itself as indicated below.

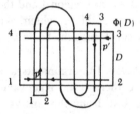

In this case there are two saddle points p and p' "fencing off" the hyperbolic set (see the beginning of Chapter 4). Below we indicate D, $\Phi(D) \setminus D$ and $\Phi^2(D) \setminus \Phi(D)$.

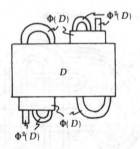

From this it is clear that the part of the plane, fenced off by the separatrices of p

and p' has the form:

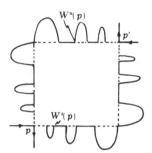

Finally we put this horseshoe in the projective plane and indicate the line l along which to push the unstable separatrix of p to its stable separatrix.

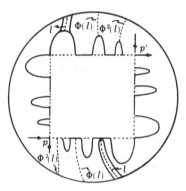

Case (c.2); p part of a non-trivial basic set.

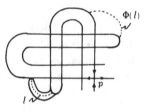

II. *Eigenvalues* $+\,-$.

Case (a.1); p isolated in $\tilde{\Omega}(\varphi)$.

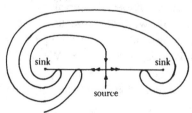

Case (a.2); p isolated in $\tilde{\Omega}(\varphi)$.

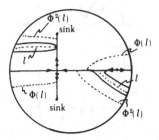

Also this is an example in the projective plane.

3. Intersections and almost-intersections of scaled sets

In this chapter we consider (almost) intersections of scaled sets (basically of Cantor type) which depend on a parameter. In the next chapter these results will be used to prove that, under certain hypotheses, the measure of the Ω-bifurcation set, after a homoclinic Ω-explosion, is very small. One can also compare the present exposition with [11]; there are several differences, mainly due to the fact that we show here that the "bifurcation set" B, as defined in the introduction, is relatively small in the sense that

$$\lim_{\mu \to 0} \frac{m(B \cap [0, \mu])}{\mu} = 0$$

(m denotes the Lebesgue measure on **R**), whereas in [11] we only proved that the lim inf of the above expression is zero.

1. *Almost intersections of rigid scaled sets.* We say that a Cantor set, or more generally a closed subset $A \subset \mathbf{R}$, is *scaled* with *scaling factor* $\lambda \in (0, 1)$ if A is invariant under scalar multiplication by λ. We recall the definition of limit capacity from the introduction. For a compact subset $C \subset \mathbf{R}$ we define the *limit capacity* $d(C)$ of C as

$$d(C) = \limsup_{\varepsilon \to 0} \frac{\text{Log } n(C, \varepsilon)}{-\text{Log } \varepsilon},$$

where $n(C, \varepsilon)$ is the minimal number of intervals of length ε needed to cover C. For a scaled subset $A \subset \mathbf{R}$, we define the limit capacity $d(A)$ as the limit capacity, in the above sense, of $A \cap [-r, +r]$ for some $r > 0$; due to the scaling, this definition is independent of r. We shall also use the following notation: If $A \subset \mathbf{R}$ is a subset, then $A^r = A \cap [-r, +r]$, A_ε is an ε-neighborhood of A and A^r_ε is the intersection of an ε-neighborhood of A with $[-r, +r]$, so that $A_\varepsilon^r = A_\varepsilon \cap [-r, +r]$. In this section we prove:

PROPOSITION. *Let $A, B \subset \mathbf{R}$ be scaled subsets of $\mathbf{R}$ with limit capacities $d(A)$ and $d(B)$ and let K be a fixed constant. For $r, \varepsilon > 0$,*

$$M_{r,\varepsilon} = \{t \in [0, r] \,|\, A_{\varepsilon \cdot r} \cap (B_{\varepsilon \cdot r} + t) \cap [-K \cdot r, K \cdot r] \neq \varnothing\}.$$

If $d(A) + d(B) < 1$, then for each $\delta > 0$ there is an $\varepsilon(\delta)$ such that $1/r \cdot m(M_{r,\varepsilon}) < \delta$ whenever $\varepsilon < \varepsilon(\delta)$; m denotes the Lebesgue measure.

Remark. One can consider values $t \in M_{r,\varepsilon}$ as the values for which the translate by t of B intersects, or *almost intersects*, A in a $(K \cdot r)$-neighborhood of 0. The constant K which may seem here somewhat artificial is forced upon us by the applications (see Chapter 4) which we have in mind.

Before we start the proof of this proposition we first state and prove the following lemma.

LEMMA. *Let A be a scaled set with limit capacity smaller than $d'(A)$. Then there is a constant a so that A^r can be covered with at most $a \cdot (\varepsilon/r)^{-d'(A)}$ intervals of length ε whenever $\varepsilon < r$.*

Proof. Since the limit capacity of A is smaller than $d'(A)$, there is a function $a(r) > 0$ of r such that A^r can be covered with at most $a(r) \cdot \varepsilon^{-d'(A)}$ intervals of length ε whenever $\varepsilon < r$. Let λ be the scaling factor of A. Because of the scaling, $A^{\lambda \cdot r}$ can be covered by $a(r) \cdot \varepsilon^{-d'(A)}$ intervals of length $\lambda \cdot \varepsilon$; on the other hand it can be covered by $a(\lambda \cdot r) \cdot (\lambda \cdot \varepsilon)^{-d'(A)}$ such intervals. So we may assume that $a(\lambda \cdot r) = \lambda^{d'(A)} \cdot a(r)$. From this, and the fact that we may assume that $a(r) = a(1)$ for $r \in (\lambda, 1]$, we conclude that there is a constant a such that $a(r) \leq a \cdot r^{d'(A)}$. This implies the lemma.

Proof of the proposition. According to the lemma there is a constant a such that $A^{(K+1) \cdot r}$ can be covered by at most

$$a \cdot \left(\frac{\varepsilon \cdot r}{(K+1) \cdot r}\right)^{-d'(A)}$$

intervals of length $\varepsilon \cdot r$ ($d'(A)$ is some number greater than $d(A)$). Hence, $A_{\varepsilon \cdot r}^{(K+1) \cdot r}$ can be covered by the same number of intervals of length $3 \cdot \varepsilon \cdot r$. In the same way, $B_{\varepsilon \cdot r}^{(K+1) \cdot r}$ can be covered by at most

$$b \cdot \left(\frac{\varepsilon \cdot r}{(K+1) \cdot r}\right)^{-d'(B)}$$

intervals of length $3 \cdot \varepsilon \cdot r$ (also $d'(B) > d(B)$, but we take $d'(A)$ and $d'(B)$ so that $d'(A) + d'(B) < 1$). This means that $M_{r,\varepsilon}$ can be covered by at most

$$a \cdot b \cdot \left(\frac{\varepsilon \cdot r}{(K + 1) \cdot r} \right)^{-d'(A) - d'(B)}$$

intervals of length $6 \cdot \varepsilon \cdot r$. From this it follows that

$$m(M_{r,\varepsilon}) \leq 6 \cdot r \cdot a \cdot b \cdot \left(\frac{1}{K + 1} \right)^{-d'(A) - d'(B)} \cdot \varepsilon^{1 - d'(A) - d'(B)}.$$

Since $d'(A) + d'(B) < 1$, $1/r \cdot m(M_{r,\varepsilon})$ goes to zero for $\varepsilon \to 0$. This proves the proposition.

2. *Continuous 1-parameter families of scaled sets.* We say that a 1-parameter family $A(\mu) \subset \mathbf{R}$, $\mu \in \mathbf{R}$, of scaled sets is *continuous* if:

(i) For each $r > 0$, the map $\mu \mapsto A(\mu) \cap [-r, +r]$ is continuous with respect to the Hausdorff metric for compact subsets of $\mathbf{R}$;

(ii) The scaling factor $\lambda(\mu)$ of $A(\mu)$ is a Lipschitz function of μ.

For such 1-parameter families $A(\mu)$ we often denote $A(0)$ simply by A.

PROPOSITION. *Let $A(\mu)$ be a continuous 1-parameter family of scaled sets and let $K > 0$ be a fixed constant. Then, for each $\varepsilon > 0$ there is an $r(\varepsilon) > 0$ such that for each $0 < r < r(\varepsilon)$,*

$$A_{r,\varepsilon} \supset \left(A(\mu) \cap [-(K + 1) \cdot r, (K + 1) \cdot r] \right)$$

for all $0 < \mu < r$.

Proof. The proof of this proposition is complicated, mainly because of the fact that the scaling factor may not be constant. If $\lambda(\mu) \equiv \lambda(0)$, then the proof is quite simple: we just take $r(\varepsilon)$ so small that the distance, in the Hausdorff metric, between $A \cap [-1, +1]$ and $A(\mu) \cap [-1, +1]$ is less than $\varepsilon/K + 1$ for all $0 < \mu < r(\varepsilon)$.

From now on we concentrate on the consequences of the fact that the scaling factor may not be constant. To be more specific, we prove the proposition for the case where $A(\mu) = \{ \lambda(\mu)^i | i \geq 0 \} \cup \{0\}$, this being the simplest case with non-constant scaling factor. The general case can then be done in the same way.

LEMMA. *Let $p(\mu) = (\lambda(\mu))^i$ and $h = p(0)$. If $\Delta\lambda$ is so small that $1/2 \leq \mathrm{Log}(\lambda(0) + \Delta\lambda)/\mathrm{Log}\,\lambda(0)$, then for $|\lambda(\mu) - \lambda(0)| < \Delta\lambda$,*

$$|p(\mu) - p(0)| \leq \frac{\mathrm{Log}\,h}{\mathrm{Log}\,\lambda(0)} \cdot \left(\frac{h}{\lambda(0)} \right)^{1/2} \cdot \Delta\lambda.$$

Proof. We define $f(\lambda) = \lambda^i$; then $f'(\lambda) = i \cdot \lambda^{i-1}$. The maximum of $f'(\lambda)$, for $\lambda \in [\lambda(0) - \Delta\lambda, \lambda(0) + \Delta\lambda]$ is attained at $\lambda(0) + \Delta\lambda$ and this maximum is $i \cdot (\lambda(0) + \Delta\lambda)^{i-1}$. From this we get $|p(\mu) - p(0)| \le i \cdot (\lambda(0) + \Delta\lambda)^{i-1} \cdot \Delta\lambda$. Since $i = \text{Log } h/\text{Log } \lambda(0)$, we find

$$|p(\mu) - p(0)| \le \frac{\text{Log } h}{\text{Log } \lambda(0)} \cdot e^{((\text{Log } h - \text{Log } \lambda(0))/\text{Log } \lambda(0)) \cdot \text{Log}(\lambda(0) + \Delta\lambda)} \cdot \Delta\lambda$$

$$\le \frac{\text{Log } h}{\text{Log } \lambda(0)} \cdot \left(\frac{h}{\lambda(0)} \right)^{1/2} \cdot \Delta\lambda$$

(for this last inequality we needed that $1/2 \le \text{Log}(\lambda(0) + \Delta\lambda)/\text{Log } \lambda(0)$). For further use, let

$$M(h, \lambda(0)) = \frac{\text{Log } h}{\text{Log } \lambda(0)} \cdot \left(\frac{h}{\lambda(0)} \right)^{1/2}.$$

LEMMA. *For fixed* $\lambda(0)$ *and* $h \in (0, 1)$ *variable, M attains its maximum for* $h = e^{-2}$; *this maximum value is* $M_{\max}(\lambda_0) < -1/\text{Log } \lambda(0) \cdot (\lambda(0))^{1/2}$.

Proof.

$$\frac{dM}{dh} = \frac{1}{h \cdot \text{Log } \lambda(0)} \cdot \left(\frac{h}{\lambda(0)} \right)^{1/2} + \frac{1}{2} \cdot \frac{\text{Log } h}{\text{Log } \lambda(0)} \cdot \left(\frac{1}{h \cdot \lambda(0)} \right)^{1/2}$$

$$= \frac{1}{\text{Log } \lambda(0)} \cdot \left(\frac{1}{h \cdot \lambda(0)} \right)^{1/2} \cdot \left(1 + \tfrac{1}{2}\text{Log } h \right).$$

This is zero when $h = e^{-2}$. Since $M(1, \lambda(0)) = 0$, $\lim_{h \searrow 0} M(h, \lambda(0)) = 0$, and $M(h, \lambda(0)) > 0$ for $h \in (0, 1)$, $M(h, \lambda(0))$ attains its maximum for $h = e^{-2}$. Then

$$M(e^{-2}, \lambda(0)) = \frac{\text{Log } e^{-2}}{\text{Log } \lambda(0)} \cdot \left(\frac{e^{-2}}{\lambda(0)} \right)^{1/2}$$

$$= \frac{-2 \cdot e^{-1}}{(\lambda(0))^{1/2} \cdot \text{Log } \lambda(0)}$$

$$< \frac{-1}{\text{Log } \lambda(0) \cdot (\lambda(0))^{1/2}}.$$

Remark. From the above calculations we see that for $r \le e^{-2}$, the maximum of $M(h, \lambda(0))$, for fixed $\lambda(0)$ and variable $h \in (0, r]$, is attained for $h = r$

and this maximum is

$$M(r, \lambda(0)) = \frac{\text{Log } r}{\text{Log } \lambda(0)} \cdot \frac{r^{1/2}}{(\lambda(0))^{1/2}} \leq r^{1/3};$$

for this last inequality we have to assume that r is sufficiently small, say $r \leq r_0$.

Now we go back to the proof of the proposition for

$$A(\mu) = \{(\lambda(\mu))^i | i \geq 0\} \cup \{0\}.$$

For some constant L, the variation in the scaling factor $\Delta\lambda$ is smaller or equal $L \cdot r(\varepsilon)$ for $0 < \mu < r(\varepsilon)$ (since $\lambda(\mu)$ is Lipschitz). Furthermore, we assume that $r(\varepsilon) \cdot (K + 1) \leq r_0$, r_0 as in the above remark. Then, from the previous results it follows that $A(\mu) \cap [- (K + 1) \cdot r, (K + 1) \cdot r]$ is contained in an $L \cdot r \cdot ((K + 1) \cdot r)^{1/3}$-neighborhood of $A(0)$. This means that, for given $\varepsilon > 0$, we have to choose $r(\varepsilon)$ so that $0 < r(\varepsilon) < r_0$ and so that $\varepsilon > L \cdot ((K + 1) \cdot r(\varepsilon))^{1/3}$. For such a value of $r(\varepsilon)$ the conclusion of the proposition holds.

3. *Diffeomorphic images of scaled sets and the final result.*

PROPOSITION. *Let $A(\mu)$ be a continuous 1-parameter family of scaled sets and let $\Psi_\mu : \mathbf{R} \to \mathbf{R}$ be a continuous 1-parameter family of C^1-diffeomorphisms with $\Psi_\mu(0) = 0$ and $\Psi_\mu'(0) = 1$. Let $K > 0$ be a constant. Then, for every $\varepsilon > 0$ there is an $r(\varepsilon) > 0$ such that*

$$A_{r \cdot \varepsilon}(\mu) \supset \left(\Psi_\mu(A(\mu)) \cap [- (K + 1) \cdot r, (K + 1) \cdot r]\right)$$

for all r, $\mu \in [0, r(\varepsilon)]$.

This is a simple consequence of the fact that for $\mu \in [0, r(\varepsilon)]$,

$$\frac{\Psi_\mu(x) - x}{|x|} \to 0$$

for $|x| \to 0$, uniformly in μ; see also [11].

In the case that $\Psi_\mu'(0) \neq 1$, one has to replace, in the conclusion, $A_{r \cdot \varepsilon}(\mu)$ by the $(r \cdot \varepsilon)$-neighborhood of $\Psi_\mu'(0) \cdot A(\mu)$.

In Section 1, we investigated the set of values of t such that A has an almost intersection with B, shifted over a distance t. In Section 2 we analyzed 1-parameter families of scaled sets depending continuously on a parameter μ. In the present section we gave an estimate on the deformation of a scaled set by a C^1-diffeomorphism depending on μ. Combining all this we obtain the following theorem.

THEOREM. *Let $A(\mu)$ and $B(\mu)$ be two continuous 1-parameter families of scaled sets so that $d(A) + d(B) < 1$. Let $\Psi_\mu, \Phi_\mu : \mathbf{R} \to \mathbf{R}$ be two continuous*

1-parameter families of C^1-diffeomorphisms such that $\Psi_\mu(0) = 0$ and $\Phi_\mu(0) = \mu$. Let $K > 0$ be a constant and let $M_{r,\,\varepsilon}$ be the set of those $\mu \in [0, r]$ such that the distance between $\Psi_\mu(A(\mu)) \cap [- K \cdot r, K \cdot r]$ and $\Phi_\mu(B(\mu)) \cap [- K \cdot r, K \cdot r]$ is smaller than $\varepsilon \cdot r$. Then, for each $\delta > 0$, there is an $\varepsilon(\delta)$ such that

$$\frac{m(M_{r,\,\varepsilon(\delta)})}{r} < \delta$$

for r sufficiently small, where m denotes Lebesgue measure.

4. Proof of the main theorem

This chapter culminates with a proposition on the hyperbolicity of the nonwandering set (Section 4), which together with a key estimate in Section 3 yield our main result as stated in the introduction.

1. *Filtrations.* The 1-parameter families φ_μ: $M \to M$ with homoclinic Ω-explosion which we consider are persistently hyperbolic for $\mu < 0$. We first want to derive some consequences from this last assumption. For a diffeomorphism φ: $M \to M$, M a closed surface, we know that if $\Omega = \Omega(\varphi)$ is hyperbolic then the periodic points are dense in Ω, where Ω denotes the nonwandering set. Then it follows, e.g. see [14], that Ω is the disjoint union of a finite number of basic sets Λ_i. Recall that Λ is a *basic set* for a diffeomorphism φ if Λ is a compact φ-invariant hyperbolic set, in which the periodic orbits are dense, and which has a dense orbit (see the introduction).

For a basic set Λ we define its *stable set* $W^s(\Lambda)$ as $\{ x \in M | \rho(\varphi^n(x), \Lambda) \to 0$ for $n \to \infty\}$, where ρ is some metric on M. $W^s(\Lambda)$ is the union of the stable sets of the points of Λ: $W^s(\Lambda) = \bigcup_{y \in \Lambda} W^s(y)$, where

$$W^s(y) = \{ x \in M | \rho(\varphi^n(x), \varphi^n(y)) \to 0 \text{ for } n \to \infty\}.$$

Recall that these stable sets of points of Λ are (locally) leaves of the stable foliation of Λ; for y a periodic saddle point, $W^s(y)$ is its stable separatrix. Unstable sets are defined in the same way except that in that case one takes limits with $n \to \infty$. In the case where M is a compact surface and Λ is a basic set, of a diffeomorphism φ: $M \to M$, of saddle type (in particular, the stable and unstable sets of points in Λ are 1-dimensional), then we know from [6] the following fact. For each point $y \in \Lambda$ such that $W^s(y)$ is at most accumulated from one side by stable sets (or leaves of the stable foliation) of other points of Λ, then y is in the stable separatrix of a periodic point $p \in \Lambda$, and hence $W^s(y) = W^s(p)$. Intuitively this means that such a stable set $W^s(\Lambda)$ is fenced by stable separatrices of periodic points. Combining this fact with the corresponding fact about the unstable sets we see that Λ is fenced by stable and unstable separatrices of periodic points.

The next point we want to mention about persistently hyperbolic diffeomorphisms $\varphi: M \to M$, M a closed 2-manifold, is that φ cannot have cycles and hence is Ω-stable [16]. To prove this, assume that φ has a *cycle*, i.e. that there are basic sets $\Lambda_1, \Lambda_2, \ldots, \Lambda_k = \Lambda_1$, $k \geq 2$, such that $W^u(\Lambda_i) \cap W^s(\Lambda_{i+1}) \neq \varnothing$ for $i = 1, \ldots, k - 1$. From [10] we know that, with an arbitrarily small perturbation of φ we obtain a new diffeomorphism $\tilde{\varphi}$ having one basic set Λ which contains basic (sub) sets $\tilde{\Lambda}_i \subsetneq \Lambda$ with $\tilde{\Lambda}_i$ close to Λ_i. This means that in a 1-parameter family of diffeomorphisms, connecting φ and $\tilde{\varphi}$, new parts of basic sets and hence new periodic points are created. This contradicts the assumption that φ is persistently hyperbolic and hence that we may assume that $\tilde{\varphi}$ as well as the diffeomorphisms between φ and $\tilde{\varphi}$ are hyperbolic.

Returning now to the 1-parameter family $\varphi_\mu: M \to M$, we recall that for $\mu = 0$ we have:

- $\varphi = \varphi_0$ is hyperbolic on $\lim_{\mu \nearrow 0} \Omega(\varphi_\mu) = \tilde{\Omega}(\varphi_0)$;
- φ has exactly one orbit of tangency between a stable and an unstable separatrix of periodic points, and this orbit of tangency is in fact an orbit of homoclinic tangency, i.e., of a tangency of the stable and unstable separatrix of the same saddle point p of $\varphi = \varphi_0$;
- the Ω-set of φ is the union of $\tilde{\Omega}(\varphi_0)$ and the orbit $\mathcal{O}$ of tangency.

Since, for $\mu < 0$, φ_μ is persistently hyperbolic, there are basic sets $\Lambda_i(\mu)$ depending continuously on μ such that $\Omega(\varphi_\mu)$ is the disjoint union of these basic sets. Since there are no cycles we may assume that

$$(*) \qquad W^u(\Lambda_i(\mu)) \cap W^s(\Lambda_l(\mu)) = \varnothing \quad \text{for all} \quad i < l \quad \text{and} \quad \mu < 0.$$

We define $\Lambda_i(0) = \lim_{\mu \nearrow 0} \Lambda_i(\mu)$, so that $\tilde{\Omega}(\varphi) = \bigcup \Lambda_i(0)$. We want to prove that also for $\mu = 0$, $(*)$ holds. Suppose not. Then there are $i < l$ so that

$$W^u(\Lambda_i(0)) \cap W^s(\Lambda_l(0)) \neq \varnothing.$$

We may assume that Λ_i and Λ_l are of saddle type. Since this intersection is empty for $\mu < 0$ and since the stable and unstable sets of these basic sets are fenced by stable and unstable separatrices of periodic points, there must be a tangency of the unstable separatrix of a periodic point in $\Lambda_i(0)$ with the stable separatrix of a periodic point in $\Lambda_l(0)$. But this extra tangency for $\mu = 0$ is against our assumptions.

In particular, all basic sets $\Lambda_i(\mu)$ are open in $\Omega(\varphi_\mu)$ for $\mu \leq 0$, except for $\Lambda_{i_0}(0)$. But, it is clear from the above that even $\Lambda_{i_0}(0)$ is the maximal φ_0-invariant set in some neighborhood of it. That is, there exists $V \supset \Lambda_{i_0}(0)$ such that $\bigcap_{n \in \mathbb{Z}} \varphi_0^n(V) = \Lambda_{i_0}(0)$. Then, from [3, §7], we conclude that $\varphi_0 | \Lambda_i(0)$ is conjugate to $\varphi_\mu | \Lambda_i(\mu)$ for all i and $\mu < 0$. Therefore, as we mentioned in the introduction, the basic sets we consider here are open in the nonwandering set or they are homeomorphic to the ones having this property. Moreover, we can even consider basic sets $\Lambda_i(\mu)$ for μ positive and small as (canonical) *continuation* of $\Lambda_i(0)$ in the following way. If we choose neighborhoods $V_i \supset \Lambda_i(0)$ in which $\Lambda_i(0)$ is the maximal φ_0-invariant set, then we take $\Lambda_i(\mu)$, μ positive and small, as the maximal invariant set for φ_μ in V_i. Again by [3, §7], we have that $\Lambda_i(\mu)$ is hyperbolic and $\varphi_\mu|\Lambda_i(\mu)$ is conjugate to $\varphi_0|\Lambda_i(0)$.

Let i_0 be the index of the basic set containing the saddle point p_μ of φ_μ whose stable and unstable separatrices make a tangency for $\mu = 0$ (we do not exclude that $\Lambda_{i_0}(\mu) = p_\mu$; in that case we call $\Lambda_{i_0}(\mu)$ a trivial basic set; otherwise we call it a non-trivial basic set). Observe that

$$W^s\big(\Lambda_{i_0}(0)\big) = W^s\big(\Lambda_{i_0}(0) \cup \mathcal{O}\big)$$

and

$$W^u\big(\Lambda_{i_0}(0)\big) = W^u\big(\Lambda_{i_0}(0) \cup \mathcal{O}\big).$$

So if we replace $\Lambda_{i_0}(0)$ by $\Lambda_{i_0}(0) \cup \mathcal{O}$, then (*) still holds.

By [14] there is a filtration $M_1 \subset M_2 \subset M$ such that
   (i) M_i is closed;
   (ii) $M_1 \subset \mathrm{int}(M_2)$;
   (iii) $\varphi(M_i) \subset \mathrm{int}(M_i)$;
   (iv) $\bigcap_{n \in \mathbf{z}} \varphi^n(M_2 - M_1) = \Lambda_{i_0}(0) \cup \mathcal{O}$.

This means that for any (small) neighborhood U of $\Lambda_{i_0}(0) \cup \mathcal{O}$ there is an $n(U)$ such that for each point $x \notin U$, $\varphi^{n(U)}(x)$ or $\varphi^{-n(U)}(x)$ is in the complement of $M_2 - \mathrm{Int}(M_1)$. Hence, for μ near zero one still has this same property for φ_μ instead of φ. This means that for any neighborhood U of $\Lambda_{i_0}(0) \cup \mathcal{O}$, there is a $\mu_0(U) > 0$ such that for any $\mu \in [0, \mu_0(U)]$, the nonwandering set $\Omega(\varphi_\mu)$ of φ_μ, intersected with $M_2 - M_1$, is contained in U. So for the rest of this chapter we may restrict our attention to such a neighborhood U, restricting μ to $[0, \mu_0(U)]$.

2. *Bounds on the region where new nonwandering points can appear.* Let again $\varphi_\mu : M \to M$ be a 1-parameter family of diffeomorphisms with a homoclinic Ω-explosion. As before, $\Omega(\varphi_\mu)$ denotes the set of nonwandering points of φ_μ. For $\mu < 0$, $\Omega(\varphi_\mu)$ is a hyperbolic set; the continuation of these hyperbolic sets, for $\mu \geq 0$, are denoted by $\tilde{\Omega}(\varphi_\mu)$. So $\Omega(\varphi_0) = \tilde{\Omega}(\varphi_0) \cup \mathcal{O}$, where $\mathcal{O}$ is the orbit of tangency; $\Omega(\varphi_0)$ and $\tilde{\Omega}(\varphi_0)$ will also be denoted by $\Omega(\varphi)$ and $\tilde{\Omega}(\varphi)$.

PROPOSITION. *Let r be a point in $\mathcal{O}$, the orbit of tangency of φ. Then for some constant K and $\mu > 0$, μ small, every orbit of $\Omega(\varphi_\mu) \setminus \tilde{\Omega}(\varphi_\mu)$ has a point in a $(K \cdot \mu)$-neighborhood of $W^u(p_\mu)$ near r. Again, p_μ denotes the saddle point of φ_μ with $\mathcal{O} \subset W^u(p_0) \cap W^s(p_0)$.*

Proof. We shall only give the proof in the case where the saddle point p has two positive eigenvalues, where the sides of tangencies and connections are as in

the figure below

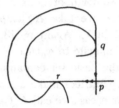

and where the expanding eigenvalue is dominating; see Chapter 2. We also assume that p is not part of a non-trivial basic set. The other cases can be treated similarly; see also the remark at the end of this section. As we have seen in the last section we can take a small neighborhood U_r of r, U_r being the component of r of a neighborhood U of $\Omega(\varphi)$, and then the "new Ω-orbits", i.e., the orbits of $\Omega(\varphi_\mu) \setminus \tilde{\Omega}(\varphi_\mu)$, will all pass through U_r and stay in U for $0 \le \mu \le \mu_0(U)$.

In the following considerations we assume that we have linearizing coordinates for φ_μ in a neighborhood of p_μ. We assume that in these coordinates $W^u_{loc}(p_\mu)$ and $W^s_{loc}(p_\mu)$ are independent of μ and that the coordinates of r and q are $(-1, 0)$ and $(0, 1)$; as usual r and q are points in the orbit of tangency in $W^u_{loc}(p_\mu)$, $W^s_{loc}(p_\mu)$. We assume that the tangency unfolds generically; to be more precise we assume the following situation near r for $\mu > 0$:

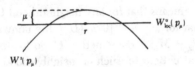

the maximum distance of $W^s(p_\mu)$ above $W^u_{loc}(p_\mu)$ is μ (here one only considers the part of $W^s(p_\mu)$ containing r). The proposition is a consequence of the following claim. For U_r and $\mu_0(U)$ as above and sufficiently small, and for any $x \in U_r$, $0 \le \mu \le \mu_0(U)$, there is a positive $n_0(x, \mu)$ such that, whenever $n' > n_0(x, \mu)$ and $\varphi_\mu^{-n'}(x) \in U_r$, then $\varphi_\mu^{-n'}(x)$ has height less than 2μ above $W^u(p_\mu)$.

To prove this claim, take a point $x \in U_r$ at distance d above $W^u_{loc}(p_\mu)$. If $\varphi_\mu^{-n}(x)$ is near q, then the distance from $\varphi_\mu^{-n}(x)$ to $W^s_{loc}(p_\mu)$ is of the order $d^{-\mathrm{Log}\,\sigma_\mu/\mathrm{Log}\,\lambda_\mu}$ where λ_μ and σ_μ are the contracting and expanding eigenvalues of $(d\varphi_\mu)_{p_\mu}$. Due to the dominance of σ, $d^{-\mathrm{Log}\,\sigma_\mu/\mathrm{Log}\,\lambda_\mu}$ is much smaller than d. We may assume that $\varphi_\mu^{-n}(x)$ is in a small neighborhood of q; otherwise it would never return to U_r under negative iterations of φ_μ (in which case the claim would be trivially true for this (x, μ)). Let m_0 be such that $\varphi^{m_0}(r) = q$. Then $\varphi_\mu^{-n-m_0}(x) \in U_r$, and the distance from $\varphi_\mu^{-n-m_0}(x)$ to $W^s(p_\mu)$ is of the order of

$d^{-\text{Log }\sigma_\mu/\text{Log }\lambda_\mu}$; i.e., there is a constant C, independent of d, such that the distance from $\varphi_\mu^{-n-m_0}(x)$ to $W^s(p_\mu)$ is smaller or equal to $C \cdot d^{-\text{Log }\sigma_\mu/\text{Log }\lambda_\mu}$. For d sufficiently small and μ near zero we have $C \cdot d^{-\text{Log }\sigma_\mu/\text{Log }\lambda_\mu} < \frac{1}{2}d$. We force d and μ to be so small that this last inequality holds by choosing U_r small. So if a point $x \in U_r$ has a negative iterate x' in U_r then $\rho(x', W^s(p_\mu)) < 1/2 \cdot \rho(x, W^u_{\text{loc}}(p_\mu))$, where ρ is the distance function corresponding to the linearizing coordinates. If x' is below $W^u_{\text{loc}}(p_\mu)$ then no further negative iterate of x will be in U_r. The maximum distance of $W^s(p_\mu)$ above $W^u_{\text{loc}}(p_\mu)$ is μ so that $\rho(x', W^u_{\text{loc}}(p_\mu)) < \mu + 1/2 \cdot \rho(x, W^u_{\text{loc}}(p_\mu))$. This implies that after returning sufficiently often to U_r under negative iteration we will finally be at a distance of less than 2μ from $W^u_{\text{loc}}(p_\mu)$. This completes the proof of the proposition in this case.

Remark. We observe that the shape of the region where the orbits of $\Omega(\varphi_\mu) \setminus \tilde{\Omega}(\varphi_\mu)$ pass near r, for $\mu > 0$, is as indicated in the figure below:

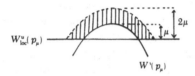

It is not hard to see that if p_μ is part of a non-trivial basic set for $\mu \leq 0$, then the points of U_r can also return (under negative iteration) *between* $W^u_{\text{loc}}(p_\mu)$ and $W^s_{\text{loc}}(p_\mu)$, $\mu > 0$, as in the case of the formation of a cycle by heteroclinic tangency; see [11].

3. *Invariant foliations and further restrictions on the locus of the non-wandering set.* We continue with the same 1-parameter family $\varphi_\mu: M \to M$ with homoclinic Ω-explosion. As before, p_μ denotes the saddle point whose stable and unstable separatrices have a tangency for $\mu = 0$. For $\mu \leq 0$, $\Lambda(\mu)$ denotes the basic set of $\Omega(\varphi_\mu)$ to which p_μ belongs. For $\mu \geq 0$, $\Lambda(\mu)$ is the basic set which is the continuation of the corresponding set for $\mu \leq 0$.

As mentioned in the introduction, there are C^1-stable and unstable foliations $\mathscr{F}_\mu^s$ and $\mathscr{F}_\mu^u$, defined on a small neighborhood of $\Lambda(\mu)$. They are C^1 in the sense that the tangent directions depend C^1 on $x \in M$, and $\mu \in R$. For x in the domain of definition of these foliations, $E_\mu^s(x)$ and $E_\mu^u(x)$ denote these tangent spaces of the leaves of $\mathscr{F}_\mu^s$ and $\mathscr{F}_\mu^u$ through x. Since $p_\mu \in \Lambda(\mu)$, $\mathscr{F}_\mu^s$ and $\mathscr{F}_\mu^u$ are both defined in a neighborhood of p_μ. As usual r and q are points in the orbit of tangency in $W^u_{\text{loc}}(p)$ and $W^s_{\text{loc}}(p)$. We extend the domain of $\mathscr{F}_\mu^u$ along $W^u(p_\mu)$ until it includes a neighborhood of r; this extension is obtained by applying some positive iterates of φ_μ to $\mathscr{F}_\mu^u$. We also extend the domain of $\mathscr{F}_\mu^s$

until it includes r, but here we use negative iterates of φ_μ; the extension of the domain proceeds along $W^s(p_\mu)$. Near r, and for small μ, there is a C^1-curve l_μ where the leaves of $\mathscr{F}_\mu^u$ and $\mathscr{F}_\mu^s$ are tangent.

In $W^s(p_\mu)$ we define the set $A(\mu)$ as $(W^s(p_\mu) \cap \Lambda(\mu)) \cup O_\mu^s$, where O_μ^s is the φ_μ-orbit of a point $q_\mu \in W^s_{loc}(p_\mu)$ depending continuously on μ and such that $q_0 = q$. In the same way $B(\mu) \subset W^u(p_\mu)$ is defined as $(W^u(p_\mu) \cap \Lambda(\mu)) \cap O_\mu^u$ where O_μ^u is the φ_μ-orbit of a point $r_\mu \in W^u(p_\mu)$ depending continuously on μ and such that $r_0 = r$. We observe that $A(\mu)$ and $B(\mu)$ are scaled sets if we identify $W^s(p_\mu)$ and $W^u(p_\mu)$ with $\mathbf{R}$ in such a way that the induced φ_μ-action becomes linear; the scaling factors are $|\lambda_\mu|$ and $|\sigma_\mu|^{-1}$. We obtain the sets $\tilde{A}(\mu)$ and $\tilde{B}(\mu)$ after projection along the leaves of $\mathscr{F}_\mu^u$ and $\mathscr{F}_\mu^s$ into l_μ.

Notice that these projections along leaves are C^1, so that $\tilde{A}(\mu)$ and $\tilde{B}(\mu)$ are like the sets $\Psi_\mu(A(\mu))$ and $\Phi_\mu(B(\mu))$ in the final result of Chapter 3. It is this situation in which the results of Chapter 3 will be applied. Before this, we still need further preparations.

Let $F_\mu \subset W^s(p_\mu)$ be a fundamental domain, i.e., a pair of two intervals in the two branches of $W^s(p_\mu)$ such that φ_μ maps one endpoint of each interval into the other endpoint, and let $\varepsilon > 0$ be a small positive number. Then there are a (small) neighborhood U of $\Lambda(0) \cup \mathcal{O}$ and a $\mu_1 > 0$, such that for each $0 < \mu < \mu_1$, whenever for some $x \in F_\mu$ the leaf of $\mathscr{F}_\mu^u$ through x contains a point of U, then there is a point x' in $A(\mu)$ such that

$$\rho(x, x') < \varepsilon \cdot \rho(p_\mu, x'),$$

where ρ measures the distance with respect to a coordinate in $W^s(p_\mu)$ which linearizes the induced φ_μ action. In the notation of [11] this means that x' is in a scaled ε-neighborhood of $A(\mu)$. We define a *scaled ε-neighborhood* of a scaled set $A \subset \mathbf{R}$ as

$$\{x \in \mathbf{R} \mid \exists\, x' \in A \text{ such that } |x - x'| < \varepsilon \cdot |x'|\};$$

such a scaled ε-neighborhood is denoted by $_\varepsilon A$ (note the difference from the notation in [11]); this should be distinguished from A_ε (see Chapter 3) which denotes an ε-neighborhood of A.

We also assume that μ_1 is so small that a neighborhood U as above exists so that for $0 < \mu < \mu_1$, $\Omega(\varphi_\mu) \setminus \tilde{\Omega}(\varphi_\mu) \subset U$ (see the last section for the notation). The component of U containing r is denoted by U_r; $\pi_{s,\mu}$ and $\pi_{u,\mu}$ denote the projections of U_r on $W^s_{loc}(p_\mu)$, $W^u_{loc}(p_\mu)$ along the leaves of $\mathscr{F}_\mu^u$ and $\mathscr{F}_\mu^s$. Then, from the fact that for $0 < \mu < \mu_1$, $\Omega(\varphi_\mu) \setminus \tilde{\Omega}(\varphi_\mu) \subset U$, it follows that all nonwandering points of φ_μ, $0 < \mu < \mu_1$ near r, are contained in

$$U_r \cap \pi_{s,\mu}^{-1}(_\varepsilon A(\mu)) \cap \pi_{u,\mu}^{-1}(_\varepsilon B(\mu))$$

(the scaled ε-neighborhoods of $A(\mu)$ and $B(\mu)$ are obtained using identifications of $W^s(p_\mu)$ and $W^u(p_\mu)$ with $\mathbf{R}$, linearizing the induced φ_μ action). In the figure below we sketch such a set.

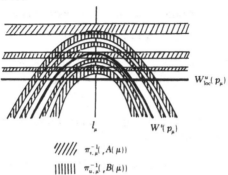

$$///// \quad \pi_{s,\mu}^{-1}(_\varepsilon A(\mu))$$

$$|||||| \quad \pi_{u,\mu}^{-1}(_\varepsilon B(\mu))$$

By use of fixed identifications of $W^u(p_\mu)$ and $W^s(p_\mu)$ with $\mathbf{R}$, it follows from the result in the previous section that we even have, for some constant K and $0 < \mu < \mu_1$,

$$\left(\Omega(\varphi_\mu) \cap U_r \right) \subset U_r \cap \pi_{s,\mu}^{-1}(_\varepsilon A(\mu) \cap [-K \cdot \mu, K \cdot \mu])$$

$$\cap \pi_{u,\mu}^{-1}(_\varepsilon B(\mu) \cap [-K \cdot \mu, K \cdot \mu]).$$

We conclude this section with an estimate on the distances between the intersections of l_μ with $\pi_{s,\mu}^{-1}(_\varepsilon A(\mu) \cap [-K \cdot \mu, K \cdot \mu])$ and with $\pi_{u,\mu}^{-1}(_\varepsilon B(\mu) \cap [-K \cdot \mu, K \cdot \mu])$, for $0 < \mu < \mu_1$; these sets are denoted by $_\varepsilon \tilde{A}^{K \cdot \mu}(\mu)$ and $_\varepsilon \tilde{B}^{K \cdot \mu}(\mu)$. It was for this estimate that the results in Chapter 3 were derived and they will be used in Section 4 to prove hyperbolicity for many values of the parameter μ.

For these estimates we need, in the above considerations, ε and μ_1 variable. The possible choices of μ_1 depend on ε and so making one choice for each ε, we have $\mu_1(\varepsilon)$; we may, and do assume that $\mu_1(\varepsilon) \leq \varepsilon$. For $0 < \mu < \mu_1(\varepsilon)$ we define

$$B_{\mu, \varepsilon} = \left\{ \mu' \in (0, \mu) \mid \text{distance between } _\varepsilon \tilde{A}^{K \cdot \mu}(\mu') \text{ and} \right.$$

$$\left. _\varepsilon \tilde{B}^{K \cdot \mu}(\mu') \text{ is smaller than } \varepsilon \cdot \mu \right\}.$$

We observe that the definition of this set is closely related to the definition of the set $M_{r, \varepsilon}$ in the final result of Chapter 3. The reason why we use "B" here is that the above defined set is strongly related to the bifurcation set of $\{\varphi_\mu\}$.

PROPOSITION. *If the limit capacities $d(A(0))$ and $d(B(0))$ of $A(0)$ and $B(0)$ have sum smaller than one, then there is for each $\delta > 0$ an $\varepsilon(\delta) > 0$ such that*

$$\frac{1}{\mu} \cdot m\left(B_{\mu, \varepsilon(\delta)} \right) < \delta$$

for μ sufficiently small; m denotes Lebesgue measure.

Proof. This follows from the final result in Chapter 3 and the following observations.

1. For scaled and non-scaled neighborhoods we have the relation

$$_\varepsilon A^r \subset A^r_{\varepsilon \cdot r}, \text{ or } _\varepsilon A \cap [-r, +r] \subset A_{\varepsilon \cdot r} \cap [-r, +r].$$

2. Because $\pi_{s,\mu}$ and $\pi_{u,\mu}$ are C^1, the ratio, between distances in $W^s_{\text{loc}}(p_\mu)$, $W^u_{\text{loc}}(p_\mu)$ and the corresponding distances in l_μ, is bounded and bounded away from zero.

4. *Hyperbolicity.* Keeping the notation of the last section we can state the result of this section as:

PROPOSITION. *For ε sufficiently small, and $\mu' \notin B_{\mu, \varepsilon}$ for some $0 < \mu' < \mu \le \mu_1(\varepsilon)$, the nonwandering set of $\varphi_{\mu'}$ is hyperbolic and even $\varphi_{\mu'}$ is persistently hyperbolic.*

Remark. As observed in the beginning of the chapter, the main theorem of this paper is a direct consequence of the proposition above together with the estimate on $m(B_{\mu, \varepsilon(\delta)})$ in the last section.

Proof. This proof goes in the same way as the corresponding proof in [11] to which we refer for the details. Here we only want to give the geometric ideas behind the proof.

As in the last section, r is a point in the orbit of tangency, U is a small neighborhood of $\Lambda(0) \cap \mathcal{O}$, whose r-component is denoted by U_r, such that each point in $\Omega(\varphi_\mu) \setminus \tilde{\Omega}(\varphi_\mu)$ has its φ_μ-orbit passing through U_r and staying in U, for μ sufficiently small. So in order to prove hyperbolicity it is enough to show that the nonwandering points in U_r are uniformly hyperbolic.

Let now $0 < \mu' < \mu \le \mu_1(\varepsilon)$, $\mu' \in B_{\mu, \varepsilon}$, be as in the theorem and let $x \in U_r \cap \Omega(\varphi_{\mu'})$. We want to show that at x the angle between $E^u_{\mu'}(x)$ and $E^s_{\mu'}(x)$ (the tangent directions to the corresponding invariant foliations; see Section 3) is at least of the order $\sqrt{\varepsilon \cdot \mu}$; in fact this angle is greater than or equal to $L \cdot \sqrt{\varepsilon \cdot \mu}$ where L is independent of x, μ, μ', and ε. This is based on the following three facts.

- If $F^u_{\mu'}(x)$ and $F^s_{\mu'}(x)$ are the leaves of $\mathcal{F}^u_{\mu'}$ and $\mathcal{F}^s_{\mu'}$ through x, then the intersections $l_{\mu'} \cap F^u_{\mu'}(x)$ and $l_{\mu'} \cap F^s_{\mu'}(x)$ have at least distance $\mu \cdot \varepsilon$; see the previous section.
- The curvature of a leaf of $\mathcal{F}^u_{\mu'}$ at x depends continuously on x and μ'; so, since we may assume that μ' is near zero and that x is near r, the curvature of leaves of $\mathcal{F}^u_{\mu'}$ in U_r is close to the curvature of $W^u(p)$ at r. A corresponding statement holds for the leaves of $\mathcal{F}^s_{\mu'}$.

– Since the tangency of $W^u(p)$ and $W^s(p)$ at r is non-degenerate, $W^u(p)$ and $W^s(p)$ have different curvatures at r.

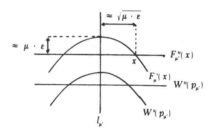

From this (see also the above figure) it follows that the distance from x to $l_{\mu'}$ is at least of the order $\sqrt{\mu \cdot \varepsilon}$. This in turn implies that the angle between $F_{\mu'}^u(x)$ and $F_{\mu'}^s(x)$ (or between $E_{\mu'}^u(x)$ and $E_{\mu'}^s(x)$) is at least of the order $\sqrt{\mu \cdot \varepsilon}$.

Next, in order to prove hyperbolicity, we construct a cone field over $\Omega(\varphi_{\mu'}) \cap U_r$; i.e., for each $x \in \Omega(\varphi_{\mu'}) \cap U_r$ we take a cone $C_{\mu'}(x) \subset T_x(M)$. The relation between such cone fields and hyperbolicity is as follows.

If, for each $x \in \Omega(\varphi_{\mu'}) \cap U_r$ and $n > 0$, such that $\varphi_{\mu'}^n(x) \in U_r$, $(d\varphi_{\mu'}^n)_x C_{\mu'}(x) \subset C_{\mu'}(\varphi_{\mu'}^n(x))$ and if $(d\varphi_{\mu'}^n)_x$ expands vectors in $C_{\mu'}(x)$ and contracts vectors in the complement of $(d\varphi_{\mu'}^{-n})_{\varphi_{\mu'}^n(x)} C_{\mu'}(\varphi_{\mu'}^n(x))$ by a factor which is exponential in n, then $\Omega(\varphi_{\mu'})$ is hyperbolic. This fact was already used in [11]. In the present situation we define $C_{\mu'}(x)$, for $x \in \Omega(\varphi_{\mu'}) \cap U_r$ to consist of those vectors in $T_x(M)$ which make an angle with $E_{\mu'}^u(x)$ smaller than half the angle between $E_{\mu'}^u(x)$ and $E_{\mu'}^s(x)$; see the following figure.

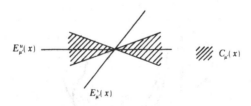

In order to see that whenever $x \in \Omega(\varphi_{\mu'}) \cap U_r$ and $\varphi_{\mu'}^n(x) \in U_r$, $n > 0$, then $(d\varphi_{\mu'}^n)_x C_{\mu'}(x) \subset C_{\mu'}(\varphi_{\mu'}^n(x))$, one has to use the fact that the orbit $\varphi_{\mu'}^i(x)$ spends a long time, namely from $i = 1$ to $i = n - 1$, in the hyperbolic region, i.e., near $p_{\mu'}$ or near the basic set containing $p_{\mu'}$. Due to this fact the successive images of $C_{\mu'}(x)$ under $(d\varphi_{\mu'}^i)_x$, $i = 0, \ldots, n$, will have the following shapes (omitting indices μ').

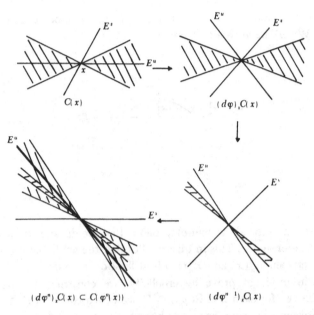

In order to justify these figures we observe that $d\varphi$ respects E^u and E^s *except* that

  – $(d\varphi)_x$ does not respect the E^u direction and

  – $(d\varphi)_{\varphi^{n-1}(x)}$ does not respect the E^s direction. This is a consequence of the construction of $\mathscr{F}^u$ and $\mathscr{F}^s$ as indicated in the figure below.

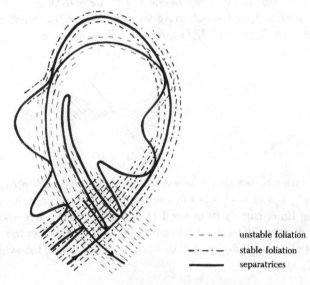

– – – –	unstable foliation
–·–·–	stable foliation
——	separatrices

(To be formally correct it may be necessary to replace, in the above description, $\varphi(x)$ by $\varphi^{i_0}(x)$ and $\varphi^{n-1}(x)$ by $\varphi^{n-i_0}(x)$, i_0 independent of x, μ, and n. See the extension of the foliation in Section 3.)

The further estimates, see [11], are based on the fact that the smallness of the angles between $E^s_\mu(x)$ and $E^u_\mu(x)$, for small μ, is compensated by the long time which the orbit spends in the hyperbolic region.

To be more precise, these estimates are based on the following facts.

- The number of iterates which such an orbit spends in the hyperbolic region (even near p_μ) is at least of the order $\mathrm{Log}\,\mu/\mathrm{Log}|\lambda|$.

- In the hyperbolic region (near p_μ), a cone, symmetric around E^u, with opening coefficient α, see the figure below, gets mapped into a cone with opening coefficient approximately $|\lambda/\sigma| \cdot \alpha$, where λ and σ are the contracting and expanding eigenvalues of $(d\varphi)_p$:

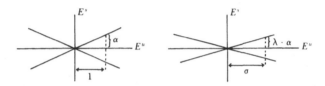

- The cone $(d\varphi_{\mu'})_x C_{\mu'}(x)$ is contained in a cone around $E^u_\mu(\varphi_{\mu'}(x))$ with opening coefficient of the order $1/\sqrt{\mu \cdot \varepsilon}$ and $\mu \leq \varepsilon$; see the last section.

This concludes our description of the proof of the hyperbolicity of $\varphi_{\mu'}$. From the proof it is also clear that $\varphi_{\mu'}$ is persistently hyperbolic.

Instituto Matemática Pura e Aplicada, Rio de Janiero, Brasil
University of Groningen, The Netherlands

References

[1] G. D. Birkhoff, Nouvelles recherches sur les systèmes dynamiques, Mem. Pont. Acad. Sci. Novi Lyncaei, 1 (1935), 85–216.

[2] P. Hartman, *Ordinary Differential Equations*, Wiley, New York, 1964.

[3] M. Hirsch and C. Pugh, Stable manifolds and hyperbolic sets, in *Global Analysis*, Proc. Symp. Pure Math., vol. 14, A. M. S., 1970, Providence, Rhode Island.

[4] J. M. Marstrand, Some fundamental geometrical properties of plane sets of fractional dimensions, Proc. London Math. Soc. (3) 4 (1954), 257–302.

[5] S. Newhouse and J. Palis, Bifurcations of Morse-Smale dynamical systems, in *Dynamical Systems*, Acad. Press, New York, 1973.

[6] ———, Hyperbolic nonwandering sets on two-dimensional manifolds, in *Dynamical Systems*, Acad. Press, New York, 1973.

[7] ———, Cycles and bifurcation theory, Astérisque, 31 (1976), 44–140.

Appendix 5

[8] S. Newhouse, The abundance of wild hyperbolic sets and non-smooth stable sets for diffeomorphisms, Publ. Math. I.H.E.S. **50** (1979), 101–151.

[9] S. Newhouse, J. Palis and F. Takens, Bifurcations and stability of families of diffeomorphisms, Publ. Math. I.H.E.S. **57** (1983), 5–71.

[10] J. Palis, A note on Ω-stability, in *Global Analysis*, Proc. Symp. Pure Math., vol. 14, A. M. S., 1970, Providence, Rhode Island.

[11] J. Palis and F. Takens, Cycles and measure of bifurcation sets for two-dimensional diffeomorphism, Inventiones Math. **82** (1985), 397–422.

[12] H. Poincaré, *Les Méthodes Nouvelles de la Mécanique Céleste*, III, Gauthiers-Villars, 1899.

[13] H. Seifert and W. Threfall, *Lehrbuch der Topologie*, Chelsea, New York, 1947.

[14] M. Shub, Stabilité globale des systèmes dynamiques, Astérisque **56** (1978), 1–211.

[15] S. Smale, Diffeomorphisms with many periodic points, in *Differential and Combinatorial Topology*, Princeton University Press, 1965.

[16] _____, The Ω-stability theorem, in *Global Analysis*, Proc. Symp. Pure Math., vol. 14, A. M. S., 1970, Providence, Rhode Island.

(Received July 15, 1985)

REFERENCES

[**A**,1961a] V.I. Arnold – On the mapping of the circle into itself, Izvestia Akad. Nauk., USSR Math. Series bf **25** (1961), 21-86.

[**A**,1961b] V.I. Arnold – Small denominators and problems of stability of motion in classical and celestial mechanics, Russ. Math. Surveys **18**, (**6**) (1963), 85-192.

[**A**,1980] V.I. Arnold – Chapitres suplémentaires de la théorie des équations différentielles ordinaires, MIR, Moscow, 1980.

[**A**,1967] D.V. Anosov – Geodesic flows on closed Riemannian manifolds with negative curvature, Proc. Stek. Inst. **90** (1967), A.M.S transl. (1969).

[**A**,1991] N. Aoki – The set of Axiom A diffeomorphisms with no cycles, Bol. Soc. Bras. Mat., to appear.

[**AP**,1937] A. Andronov and L. Pontryagin – Systèmes grossiers, Dokl. Akad. Nauk. USSR **14** (1937), 247-251.

[**AP**,1987] V.S. Afraimovich and Ya.B. Pesin – Math. Phys. Rev., Section C, **6** (1987), 169-241.

[**AM**,1991] A. Araujo and R. Mañé – On the existence of hyperbolic attractors and homoclinic tangencies for surface diffeomorphisms, to appear.

[**ACT**,1981] A. Arneodo, P. Coullet and C. Tresser – A possible new mechanism for the on set of turbulence, Phys. Lett. **81 A** (1981), 197-201.

[**B**,1935] G.D. Birkhoff – Nouvelles recherches sur les systèmes dynamiques, Mem. Pont. Acad. Sci. Novi. Lyncaei **1** (1935), 85-216.

[**B**,1975a] R. Bowen – Equilibrium states and the ergodic theory of Anosov diffeomorphisms, Lecture Notes in Math. **470** (1975), Springer-Verlag.

[**B**,1975b] R. Bowen – A horseshoe with positive measure, Inventiones Math. **29** (1975), 203-204.

[**B**,1977] R. Bowen – On Axiom A diffeomorphisms, Conference Board Math. Sci. **33**, A.M.S, 1977.

[**BR**,1975] R. Bowen and D. Ruelle – The ergodic theory of Axiom A flows, Inventiones Math. **29** (1975), 181-202.

[**BF**,1973] L. Block and J. Franke – Existence of periodic points for maps of S^1, Inventiones Math. **22** (1973), 69-73.

[**BS**,1979] A.A. Bunimovich and Ya.G. Sinai – Stochastic attractors in the Lorenz model, Nonlinear Waves, Nauka (1979), 212-226.

[**BC**,1985] M. Benedicks and L. Carleson – On iterations of $1 - ax^2$ on $(-1,1)$, Annals of Math. **122** (1985), 1-24.

[**BC**,1991] M. Benedicks and L. Carleson – The dynamics of the Hénon map, Annals of Math. **133** (1991), 73-169.

[**BY**,1991] M. Benedicks and L.S. Young – SBR measures for certain Hénon maps, to appear.

[**BHTB**,1990] H.W. Broer, G.B. Huitema, F. Takens and B.L.J. Braskma – Unfoldings and bifurcations of quasi-periodic tori, Memoirs of the A.M.S. **83, n. 421** (1990).

[**BLMP**,1991] R. Bamón, R. Labarca, R. Mañé and M.J. Pacífico – The explosion of singular cycles, to appear.

[**C**,1978] C. Conley – Isolated invariant sets and the Morse index, Conference Board Math. Sci. **38**, A.M.S, 1978.

[**C**,1991] M. Carvalho – Bowen–Ruelle–Sinai measures for n-dimensional DA diffeomorphisms, IMPA thesis and to appear.

[**CL**,1945] M.L. Cartwright and J.E. Littlewood – On nonlinear differential equations of the second order: I. The equation $\ddot{y} - k(1 - y^2)\dot{y} + y = b\lambda k \cos(\lambda t + u)$, k large, J. London Math. Soc. **20** (1945), 180-189.

[**CT**,1978] P. Coullet and C. Tresser – Iterations d'endomorphismes et groupe de renormalisation, C. R. Acad. Sci. Paris **287** (1978), 577-580.

[**CE**,1980] P. Collet and J.-P. Eckmann – Iterated maps on the interval as dynamical systems, Birkhäuser, 1980.

[**D**,1991] L. Diaz – Persistence of nonhyperbolicity and heterodimensional cycles, IMPA thesis and to appear.

[DR,1991] L. Diaz and J. Rocha – Nonconnected heterodimensional cycles: bifurcations and stability, to appear.

[DRV,1991] L. Diaz, J. Rocha and M. Viana – Saddle-node critical cycles and prevalence of strange attractors, to appear.

[ER,1985] J.-P. Eckmann and D. Ruelle – Ergodic theory of chaos and strange attractors, Rev. Mod. Phys. **57** (1985), 617-656.

[F,1978] M. J. Feigenbaum – Qualitative universality for a class of nonlinear transformations, J. Stat. Phys. **19** (1978), 25-52.

[F,1979] M. J. Feigenbaum – The universal metric properties of nonlinear transformations, J. Stat. Phys. **21** (1979), 669-709.

[F,1985] K. J. Falconer – The geometry of fractal sets, Cambridge Univ. Press, 1985.

[F,1985] J. Franks – Period doubling and the Lefschetz formula, Trans. A.M.S. **287** (1985), 275-283.

[FS,1977] J. Franke and J. Selgrade – Hyperbolicity and chain recurrence, Trans. A.M.S. **245** (1978), 251-262.

[GW,1979] J. Guckenheimer and R. F. Williams – Structural stability of Lorenz attractors, Publ. Math. IHES. **50** (1979), 59-72.

[GH,1983] J. Guckenheimer and P. Holmes – Nonlinear oscillations, dynamical systems and bifurcations of vector fields, Springer-Verlag, 1983.

[GST,1989] J.-M. Gambaudo, S. von Strein and C. Tresser – Hénon-like maps with strange attractors: there exist C^∞ Kupka–Smale diffeomorphisms on S^2 with neither sinks or sources, Nonlinearity **2** (1989), 287-304.

[H,1947] M. Hall – On the sum and product of continued fractions, Annals of Math. **48** (1947), 966-993.

[H,1964] P. Hartman – Ordinary differential equations, Wiley, 1964.

[H,1976] M. Hénon – A two-dimensional mapping with a strange attractor, Comm. Math. Phys. **50** (1976), 69-77.

[H,1977] M. Herman – Mésure de Lebesgue et nombre de rotation, Geometry and Topology, ed. J. Palis and M. do Carmo,

226   *References*

Lecture Notes in Math. **597** (1978), 271-293, Springer-Verlag.

[**H**,1979] M. Herman – Sur la conjugaison différentiable des difféomorphisms du cercle à des rotations, Publ. Math. **49** (1979), 5-233.

[**H**,1991] S. Hayashi – Diffeomorphisms in $\mathcal{F}^1(M)$ satisfy Axiom A, Erg. Th. and Dyn. Syst., to appear.

[**HP**,1970] M. Hirsch and C. Pugh – Stable manifolds and hyperbolic sets, Proc. A.M.S. Symp. Pure Math. **14** (1970), 133-163.

[**HPPS**,1970] M. Hirsch, J. Palis, C. Pugh and M. Shub – Neighbourhoods of hyperbolic sets, Inventiones Math. **9** (1970), 121-134.

[**HPS**,1977] M. Hirsch, C. Pugh and M. Shub – Invariant manifolds, Lecture Notes in Math. **583** (1977), Springer-Verlag.

[**J**,1971] M. Jacobson – On smooth mappings of the circle into itself, Math. USSR Sb. **14** (1971), 161-185.

[**J**,1981] M. Jacobson – Absolutely continuous invariant measures for one parameter families of one-dimensional maps, Comm. Math. Phys. **81** (1981), 39-88.

[**K**,1957] A.N. Kolmogorov – General theory of dynamical systems and classical mechanics, Proc. Int. Congress of Math. 1954, 315-333, North Holland.

[**K**,1964] I. Kupka – Contribution à la théorie des champs génériques, Cont. Diff. Equ. **2** (1963), 457-484, **3** (1964), 411-420.

[**K**,1980] A. Katok – Lyapunov exponents, entropy and periodic orbits for diffeomorphisms, Publ. Math. IHES **51** (1980), 137-174.

[**L**,1949a] N. Levinson – A second order differential equation with singular solutions, Annals of Math. **50** (1949), 127-153.

[**L**,1957a] J.E. Littlewood – On non-linear differential equations of second order: III, Acta Math. **97** (1957), 267-308.

[**L**,1957b] J.E. Littlewood – On non-linear differential equations of second order: IV, Acta Math. **98** (1957), 1-110.

[**L**,1963] E.N. Lorenz – Deterministic non-periodic flow, J. Atmos. Sci. **20** (1963), 130-141.

[**L**,1980] S.T. Liao – On the stability conjecture, Chinese Annals of Math. **1** (1980), 9-30.

[**L**,1981] M. Levi – Qualitative analysis of the periodically forced relaxation oscillations, Memoirs of the A.M.S. **32, n. 244** (1981).

[**LP**,1986] R. Labarca and M.J. Pacífico – Stability of singular horseshoe, Topology **25** (1986), 337-352.

[**L**,1990] P. Larsson – L'ensemble différence de deux ensembles de Cantor aléatoires, C.R.A.S. Paris **310** (1990), 735-738.

[**M**,1954] J.M. Marstrand – Some fundamental properties of plane sets of fractional dimensions, Proc. London Math. Soc. **4** (1954), 257-302.

[**M**,1962] J. Moser – On invariant curves of area preserving mappings of an annulus, Nachr. Akad. Wiss. Göttingen, Math. Phys. Kl (1962), 1-20.

[**M**,1967] J. Moser – Convergent series expansions for quasi-periodic motions, Math. Annalen. **169** (1967), 136-176.

[**M**,1969] J. Moser – On a theorem of Anosov, J. Diff. Equ. **5** (1969), 411-440.

[**M**,1973a] J. Moser – Stable and random motions in dynamical systems, Annals of Math. Studies, Princeton Univ. Press, 1973.

[**M**,1973b] W. de Melo – Structural stability of diffeomorphisms on two-manifolds, Inventiones Math. **21** (1973), 233-246.

[**M**,1978] R. Mañé – Contribution to the stability conjecture, Topology **17** (1978), 383-396.

[**M**,1982] R. Mañé – An ergodic closing lemma, Annals of Math. **116** (1982), 503-540.

[**M**,1987] R. Mañé – Ergodic theory and differentiable dynamics, Springer-Verlag, 1987.

[**M**,1988] R. Mañé – A proof of the C^1 stability conjecture, Publ. Math. I.H.E.S. **66** (1988), 161-210.

[**M**,1985] J. Milnor – On the concept of attractor, Comm. Math. Phys. **99** (1985), 177-195.

[**M**,1991] J.C. Martín-Rivas – Homoclinic bifurcations and cascades of period doubling bifurcations in higher dimensions, IMPA thesis and to appear.

[**MS**,1980] M. Misiurewicz and B. Szewc – Existence of a homoclinic orbit for the Hénon map, Comm. Math. Phys. **75** (1980), 285-291.

[**MM**,1983] A. Manning and H. McCluskey – Hausdorff dimension for horseshoes, Erg. Th. and Dyn. Syst. **3** (1983), 251-261.

[**MV**,1991] L. Mora and M. Viana – Abundance of strange attractors, Acta Math., to appear.

[**N**,1970] S. Newhouse – Non-density of Axiom A(a) on S^2, Proc. A.M.S. Symp. Pure Math. **14** (1970), 191-202.

[**N**,1972] S. Newhouse – Hyperbolic limit sets, Trans. A.M.S. **167** (1972), 125-150.

[**N**,1974] S. Newhouse – Diffeomorphisms with infinitely many sinks, Topology **13** (1974), 9-18.

[**N**,1979] S. Newhouse – The abundance of wild hyperbolic sets and nonsmooth stable sets for diffeomorphisms, Publ. Math. I.H.E.S. **50** (1979), 101-151.

[**N**,1980] S. Newhouse – Lectures on dynamical systems, in: J. Guckenheimer, J. Moser, S. Newhouse, Dynamical systems, CIME Lectures – Bressanone, Birkhäuser, 1980.

[**NP**,1973] S. Newhouse and J. Palis – Hyperbolic nonwandering sets on two-manifolds, Dynamical systems, ed. M. Peixoto, Acad. Press, 1973, 293-301.

[**NP**,1976] S. Newhouse and J. Palis – Cycles and bifurcation theory, Astérisque **31** (1976), 44-140.

[**NPT**,1983] S. Newhouse, J. Palis and F. Takens – Bifurcations and stability of families of diffeomorphisms, Publ. Math. I.H.E.S. **57** (1983), 5-71.

[**O**,1968] V. Oseledec – A multiplicative ergodic theorem: Lyapunov characteristic numbers for dynamical systems, Trans. Moscow Math. Soc. **19** (1968), 197-231.

[**P**,1890] H. Poincaré – Sur le problème des trois corps et les équations de la dynamique (Mémoire couronné du prise de S.M. le roi Oscar II de Suède), Acta Math. **13** (1890), 1-270.

[**P**,1920] B. van der Pol – De amplitude van vrije en gedwongen triode-trillingen, Tijdschr. Ned. Radiogenoot. **1** (1920), 3-31.

[**P**,1934] B. van der Pol – The nonlinear theory of electric oscillations, Proc. of the Inst. of Radio Eng. **22** (1934), 1051-1086; reprinted in: Selected scientific papers, North-Holland, 1960.

[**P**,1960] V. Pliss – Sur la grossièreté des équations différentielles définies sur le tore, Vestnik LGU, ser. mat. **13** (1960), 15-23.

[**P**,1962] M. Peixoto – Structural stability on two-manifolds, Topology **1** (1962), 101-120.

[**P**,1967] C. Pugh – The closing lemma, Amer. J. Math. **89** (1967), 956-1009.

[**P**,1969] J. Palis – On Morse Smale dynamical systems, Topology **8** (1969), 385-405.

[**P**,1970] J. Palis – A note on Ω-stability, Proc. A.M.S. Symp. Pure Math. **14** (1970), 221-222.

[**P**,1988] J. Palis, On the C^1 Ω-stability conjecture, Publ. Math. I.H.E.S. **66** (1988), 211-215.

[**PM**,1982] J. Palis and W. de Melo – Geometric theory of dynamical systems, Springer-Verlag, 1982.

[**PS**,1970] J. Palis and S. Smale – Structural stability theorems, Proc. A.M.S. Symp. Pure Math. **14** (1970), 223-232.

[**PT**,1985] J. Palis and F. Takens – Cycles and measure of bifurcation sets for two-dimensional diffeomorphims, Inventiones Math. **82** (1985), 397-422.

[**PT**,1987] J. Palis and F. Takens – Hyperbolicity and the creation of homoclinic orbits, Annals of Math. **125** (1987), 337-374. Reproduced in Appendix 5 in this book.

[**PV**,1988] J. Palis and M. Viana – Continuity of Hausdorff dimension and limit capacity for horseshoes, Dynamical Systems, Lecture Notes in Math. **1331**, (1988), 150-160.

[**PV**,1991] J. Palis and M. Viana – High dimension diffeomorphisms displaying infinitely many sinks, to appear.

[**PY**,1991] J. Palis and J. C. Yoccoz – Homoclinic tangencies for hyperbolic sets of large Hausdorff dimension, to appear.

[**P**,1974] R. V. Plykin – Sources and currents of A-diffeomorphisms of surfaces, Math. USSR Sbornik **94** (1974), 2, 243-264.

[P,1976] Ya. Pesin – Families of invariant manifolds corresponding to non-zero characteristic exponents, Math. USSR Izvestjia **10** (1976), 1261-1305.

[P,1977] Ya. Pesin – Characteristic Lyapunov exponents and ergodic theory, Russian Math. Surveys **32** (1977), 55-114.

[P,1983] D. Pixton, Markov neighbourhoods for zero - dimensional basic sets, Trans. A.M.S. **279** (1983), 431-462.

[R,1971] J. Robbin – A structural stability theorem, Annals of Math. **94** (1971), 447-493.

[R,1973] C. Robinson – C^r structural stability implies Kupka–Smale, Dynamical Systems, ed. M. Peixoto, Acad. Press, 1973, 443-449.

[R,1974] C. Robinson – Structural stability of vector fields, Annals of Math. **99** (1974), 154-175.

[R,1976] C. Robinson – Structural stability of C^1 diffeomorphisms, J. Diff. Equ. **22** (1976), 28-73.

[R,1983] C. Robinson – Bifurcation to infinitely many sinks, Comm. Math. Phys. **90** (1983), 433-459.

[R,1989] C. Robinson – Homoclinic bifurcation to a transitive attractor of Lorenz type, Nonlinearity **2** (1989), 495-518.

[R,1976] D. Ruelle – A measure associated with Axiom A attractors, Amer. J. Math. **98** (1976), 619-654.

[R,1978] D. Ruelle – An inequality of the entropy of differentiable maps, Bol. Soc. Bras. Mat. **9** (1978), 83-87.

[RT,1971] D. Ruelle and F. Takens – Comm. Math. Phys. **20** (1971), 167-192 and **23** (1971), 343-344.

[R,1986] M. Rees – Positive measure sets of ergodic rational maps, Ann. Scient. Éc. Norm. Sup. **19** (1986), 383-407.

[R,1989] M. Rychlik – Lorenz's attractors through Silnikov-type bifurcation, Erg. Th. and Dyn. Syst. **10** (1990), 793-821.

[R,1992] A. Rovella – The dynamics of the perturbations of the contracting Lorenz attractor, IMPA thesis and to appear.

[R,1992] N. Romero – Persistence of homoclinic tangencies in higher dimensions, IMPA thesis and to appear.

[S,1957] S. Sternberg – Local contractions and a theorem of Poincaré, Amer. J. Math. **79** (1957), 809-824.

[**S**,1958] S. Sternberg – On the structure of local homeomorphisms of Euclidean n-space, II, Amer. J. Math. **80** (1958), 623-631.

[**S**,1963] S. Smale – Stable manifolds for differential equations and diffeomorphisms, Ann. Scuola Sup. Pisa **17** (1963), 97-116.

[**S**,1965] S. Smale – Diffeomorphisms with many periodic points, Diff. and Comb. Topology, Princeton Univ. Press (1965), 63-80.

[**S**,1967] S. Smale – Differentiable dynamical systems, Bull. A.M.S. **73** (1967), 747-817.

[**S**,1970] S. Smale – The Ω–stability theorem, Proc. A.M.S. Symp. Pure Math. **14** (1970), 289-297.

[**S**,1965] L.P. Silnikov – A case of the existence of a denumerable set of periodic motions, Sov. Math. Dokl. **6** (1965), 163-166.

[**S**,1968] Ya. Sinai – Markov partitions and C-diffeomorphisms, Func. Anal. and its Appl. **2** (1968), 64-89.

[**S**,1970] Ya. Sinai – Dynamical systems with elastic reflections: properties of dispersing billiards, Russ. Math. Surveys **25** (1970), 137-189.

[**S**,1976] Ya. Sinai – Introduction to ergodic theory, Princeton Univ. Press, 1976.

[**S**,1971] M. Shub – Topological transitive diffeomorphisms on T^4, Lecture Notes in Math. **206** (1971), 39, Springer-Verlag.

[**S**,1978] M. Shub – Stabilité global des systèmes dynamiques, Astérisque **56** (1978).

[**S**,1979] S.J. van Strien – Centre manifolds are not C^∞, Math. Z. **166** (1979), 143-145.

[**S**,1981] S.J. van Strien – On the bifurcations creating horseshoes, Dynamical Systems, Lecture Notes in Math. **898** (1981), 316-351, Springer-Verlag.

[**S**,1983] A. Sannami – The stability theorems for discrete dynamical systems on two-dimensional manifolds, Nagoya Math. J. **90** (1983), 1-55.

[S,1991] A. Sannami – An example of a regular Cantor set whose difference is a Cantor set with positive measure, Hokkaido Math. J., to appear.

[SM,1971] C.L. Siegel and J. Moser – Lectures on celestial mechanics, Springer-Verlag, 1971.

[T,1988] F. Takens – Limit capacity and Hausdorff dimension of dynamically defined Cantor sets, Dynamical Systems, Lecture Notes in Math. **1331** (1988), 196-212, Springer-Verlag.

[T,1991a] F. Takens – Homoclinic bifurcations, Proc. Int. Congress of Math., Berkeley (1986), 1229-1236.

[T,1991b] F. Takens – On the geometry of non-transversal intersections of invariant manifolds and scaling properties of bifurcation sets, Pitman Research Notes in Math. Series, to appear.

[T,1992] F. Takens – Abundance of generic homoclinic tangencies in real-analytic diffeomorphisms, Bol. Soc. Bras. Mat. (1992), to appear.

[TY,1986] L. Tedeschini-Lalli and J.A. Yorke – How often do simple dynamical processes have infinitely many coexisting sinks? Comm. Math. Phys. **106** (1986), 635-657.

[U,1992] R. Ures – Approximating a Hénon-like strange attractor by a homoclinic tangency and an attracting cycle, IMPA thesis and to appear.

[V,1991a] E. Vargas – Bifurcation frequency for unimodal maps, Comm. Math. Phys., **141** (1991), 633-650.

[V,1991b] M. Viana – Strange attractors in higher dimensions, IMPA thesis and to appear.

[W,1970] R.F. Williams – The DA maps of Smale and structural stability, Proc. A.M.S. Symp. Pure Math. **14** (1970), 329-334.

[W,1979] R.F. Williams – The structure of Lorenz attractors, Publ. Math. I.H.E.S. **50** (1979), 101-152.

[W,1982] P. Walters – An introduction to ergodic theory, Springer-Verlag, 1982.

[YA,1983] J. A. Yorke and K. T. Alligood – Cascades of period doubling bifurcations: a prerequisite for horseshoes, Bull. A.M.S. **9** (1983), 319-322.

INDEX

attractor, strange 9, 138
attractor, Hénon-like 141, 143
Axiom A 3, 6

basic set 3, 7, 92, 169
basic sets and negative
  eigenvalues 347
bifurcation set 338
bounded distorsion property 58
bridge 61

Cantor set, affine 57
Cantor set, dynamically defined
  56, 58, 75, 80, 85
Cantor set, generalized affine 57
Cantor set, mid-α 57
chaos 8
chaos, apparent 34
conjugacy 2
connection 342
curve of tangencies 98
cycle 5
cycle, heteroclinic 92

denseness 62, 79, 85
denseness, local 85
diffeomorphism, Anosov 4
diffeomorphism, Morse-Smale 4
difference 62
dissipative 29, 34, 101

entropy, topological 144
expanding structure 54
expansive 4

filtration 92
first bifurcation 99, 102
foliation, stable, unstable 26, 27,
  52, 98, 162, 166, 338

gap 61, 63
generic unfolding (of a
  homoclinic tangency) 35
graph transform 154

Hausdorff α-measure 60
Hausdorff dimension 53, 60, 64,
  67, 77, 79, 85, 111
Hausdorff dimension, stable or
  unstable 92, 99, 101
homoclinic 11, 18
homoclinic bifurcation or
  homoclinically bifurcating 15,
  134
homoclinic point 337
homoclinic, primary 18
homoclinic (heteroclinic)
  tangency 7, 341
homoclinic Ω-explosion 339
horseshoe 14
hyperbolic(ity) 1, 92, 102, 339
hyperbolic set 156, 157, 338

infinitely many circles 346

lemma, λ- 155
limit capacity 61, 62, 67, 85,
  339, 358
limit capacity, stable or unstable
  92, 99, 101, 109, 110, 111, 339
limit set, positive, negative 2,
  107
limit set, hyperbolic 93
linearizing coordinates 163
Lyapunov exponent 145

manifold, (generalized) (local)
  (un)stable 154, 155, 158, 181
manifold, centre-((un)stable)
  160
Markov partitions 55, 171
maximal invariant set 92
measure of bifurcation set 111
mixing, topologically 169

nonhyperbolic diffeomorphisms
  137
nonwandering 2, 337
normally hyperbolic invariant
  manifold 160

orbit, chaotic, sensitive 8, 16, 30
orientability 348

pendulum 12
period doubling 39
persistently chaotic (or
  sensitive), fully 9
persistent (homoclinic)
  tangencies 113, 115
persistently hyperbolic 339
premature creation of horseshoe
  344
premature tangency 343
presentation 62

renormalization 34

r-normally hyperbolic 161

saddle-node 38
saddle-node critical cycle 136
SBR measure 147
scaling, scaled set 54, 358
shift 28
sides of tangency 342
stability, structural 1, 2, 111
stability, Ω- 6, 111
stable, homoclinically 134

thickness 61, 63, 77, 85
thickness, local 85
thickness, stable or unstable 113

Printed in the United States
By Bookmasters